HEALING EARTH

HEALING EARTH

*An Ecologist's Journey of Innovation
and Environmental Stewardship*

JOHN TODD

North Atlantic Books
Berkeley, California

Copyright © 2019 by John Todd. All rights reserved. No portion of this book, except for brief review, may be reproduced, stored in a retrieval system, or transmitted in any form or by any means—electronic, mechanical, photocopying, recording, or otherwise—without the written permission of the publisher. For information contact North Atlantic Books.

Published by
North Atlantic Books
Berkeley, California

Cover photo of Fjardargljufur Canyon, Iceland, by Ratnakorn Piyasirisorost
Cover design by Nicole Hayward
Book design by Happenstance Type-O-Rama

Printed in the United States of America

Healing Earth: An Ecologist's Journey of Innovation and Environmental Stewardship is sponsored and published by the Society for the Study of Native Arts and Sciences (dba North Atlantic Books), an educational nonprofit based in Berkeley, California, that collaborates with partners to develop cross-cultural perspectives, nurture holistic views of art, science, the humanities, and healing, and seed personal and global transformation by publishing work on the relationship of body, spirit, and nature.

North Atlantic Books' publications are available through most bookstores. For further information, visit our website at www.northatlanticbooks.com or call 800-733-3000.

Library of Congress Cataloging-in-Publication Data

Names: Todd, Nancy Jack, 1938- author.
Title: Healing earth : an ecologist's journey of innovation and environmental stewardship / John Todd.
Description: Berkeley, California : North Atlantic Books, [2019] | Includes index.
Identifiers: LCCN 2018030539 (print) | LCCN 2018044967 (ebook) | ISBN 9781623172992 (e-book) | ISBN 9781623172985 (pbk.)
Subjects: LCSH: Ecological engineering. | Environmental responsibility. | Environmental protection.
Classification: LCC GE350 (ebook) | LCC GE350 .T63 2019 (print) | DDC 628—dc23
LC record available at https://lccn.loc.gov/2018030539

1 2 3 4 5 6 7 8 9 VERSA 23 22 21 20 19

Printed on recycled paper

North Atlantic Books is committed to the protection of our environment. We partner with FSC-certified printers using soy-based inks and print on recycled paper whenever possible.

To my family

To Nancy, my partner and colleague in our shared life's work. She has been my most diligent editor and inspired in me her love of language.

To our three children, Rebecca, Jonathan, and Susannah. They have been a source of support and pure pleasure.

To our grandchildren: Rebecca's sons Max and Gareth; Jonathan's daughter, Anissa, and sons Roberto and Quinn; and Susannah's son, Logan. All of them have brought me joy.

The rule of no realm is mine....
But all worthy things that are in peril
as the world now stands, those are my care.
And for my part, I shall not wholly fail in my task ...
if anything passes through this night
that can still grow fair or bear fruit
and flower again in days to come.
For I too am a steward. Did you not know?

—J. R. R. TOLKIEN

CONTENTS

Foreword by Janine Benyus **ix**

CHAPTER 1:
A Vision of Hope
1

CHAPTER 2:
The Birth of an Ecological Technology
7

CHAPTER 3:
Steps to a Theory of Ecological Design
15

CHAPTER 4:
The Edge of the Sea
31

CHAPTER 5:
Restoring Polluted Waters Ecologically
43

CHAPTER 6:
Healing Degraded Streams and Rivers
51

CHAPTER 7:
The Early Evolution of Restorer Eco-Technologies
61

CHAPTER 8:
Lessons from the Sea
69

CHAPTER 9:
A Plan to Heal Marine Bays and Salt Ponds
along the Atlantic Coast of North America
79

CHAPTER 10:
Caribbean Futures
87

CHAPTER 11:
Solar- and Wind-Powered Workboats
95

CHAPTER 12:
Cleaning Up an Ongoing Oil Spill with Eco-Machines
103

CHAPTER 13:
Ocean Restorers: Ecological Hope Ships
for Marine Pollution Reduction
113

CHAPTER 14:
Designs for Southern Africa
121

CHAPTER 15:
Appalachian Spring
137

CHAPTER 16:
Re-Greening the Earth: The Challenge of the Sinai Desert
155

Notes **171**
Bibliography **173**
Acknowledgments **175**
Index **177**
About the Author **183**

FOREWORD

It behooves one to pay close attention to the life history of organisms including the fungi, plants, animals, and even the microorganisms. For me natural history is not an old-fashioned form of knowing. It is the narratives of living entities that provide the alphabet of the design vocabulary.

—JOHN TODD, FROM CHAPTER 4 IN THIS BOOK

Something troubling has occurred in the biological sciences over the last few decades, something Rutgers University professor David Ehrenfeld calls "the forgetting." The deeper we delve into life at the molecular level, the more we seem to lose touch with the mystery of the whole organism and the community to which it belongs. I have friends who spend their lives decoding the intricate genome of a plant called *Arabidopsis*. As the mouse-model of botany, and the first plant to have its genome sequenced, *Arabidopsis* is now the most intensively researched plant on the planet. I like to watch for its elevated stalk, pale flowers, and compact rosette when hiking with my genomic friends. "Here it is!" I exclaim, eager to watch them marvel at their mentor in the wild. My heart sinks when they ask: "What species is that?"

Estrangement from the whole makes it hard to appreciate, much less emulate, the systems that keep our planet humming. That's why it's important to hear from John Todd, a consummate student of the whole and a creator of ecological technologies that renew water, soils, air, and, ultimately, human communities. He is an apprentice of ecosystems, and after a lifetime of immersing himself in eelgrass meadows, oak-hickory forests, vernal ponds, and salt marshes, he gifts us with a book of operating instructions—universal patterns and principles that allow these ecologies to design, organize, and, most importantly, "re-member" themselves after insult.

Few people I know have a deeper regard for the wisdom of the natural world, or a bigger heart for the human species. John believes that an ecological renaissance is afoot, and that humans will one day participate in the grand planetary cycles as beneficial contributors. It's not that he is unaware of our current role in the great unraveling. In fact, some of his most profound stories are about discovering leached carcinogens near the homes of two friends lost to cancer, or watching the desecration of landscapes he loved "like companions." But rather than shrink from the damage, John wades into the worst and wonders how he can help.

His first instinct is to ask *where* he might find inspiration, meaning which living ecosystems will best teach him. Finding a wild system that is working beautifully, he climbs, hikes, or snorkels within it, taking a master class in the complex science of healing through relationships. Even during his watery childhood, as a small boy on the edge of a Great Lake, John studied the difference between two streams that accompanied him on his walk to school. One was crystal clear and brimming with life, the other erratic and muddy, stripped of its life by upstream golf courses and housing developments. He trained himself to see how the wild stream managed to run clear year in and year out. It had everything to do with the diversity of life forms in and around it, the cast of characters that define a place and keep it vibrant. Life connected to life can heal anything, says John, even the emergencies of our own making.

I share his belief in life's genius, and have dedicated my own life to the discipline of biomimicry, which I define as innovation inspired by nature. Surrounding oneself with life's solutions is like living in a cornucopia of what works. When I encourage people to go beyond the mimicry of just one species or one part of one species, I point to John's work as an *ecomimic*—an innovator who emulates the whole. In this book, you will have a rare opportunity to learn a checklist, still evolving, of the best practices that John has discovered over a lifetime of creating linked healing systems. I found myself highlighting page after page, thrilled to see this one-of-a-kind wisdom gathered and shared.

One of my favorite memories was watching John share his wisdom with a very tense room of exasperated foreign aid workers. We were in Boulder,

Colorado, at a design workshop to bring sustainable solutions to refugee camp settings. The Rocky Mountain Institute was hosting along with the Navy, whose leaders were seeing humanitarian missions all over the world become the new military theater as political and climate change refugees fled from brutality, droughts, floods, and fear. People were living for years, even decades, in camps that were designed to last months. How could these not-so-temporary settlements become less wasteful of human spirits and the environment?

We covered a plethora of sustainable technologies such as LED lights and smart cards, but the people most experienced with these crises grew increasingly frustrated, until one day, the whole thing blew up. Veterans of dozens of encampments challenged us to see through their eyes: "People are locusts," they said, referring to the damage hundreds of thousands of people can have. "People are a plague, and in their desperation, they ruin landscapes and no country wants to have them."

One after another, the beleaguered camp workers stood and testified to this fact. A man next to me spoke with such passion I could feel him shaking. At that point, Amory Lovins, the chairman of the Rocky Mountain Institute, pointed to John and me and asked what might be possible if we thought about this differently. What would a biologist say?

John and I were the only biologists in the room of nearly a hundred people. Haltingly, we began to offer an alternate vision. What if one day, countries actually *wanted* refugees to come because months and years of human energy and goodwill could be directed toward healing ravaged landscapes? What if the design *intent* was to create a National Park once the tents were rolled away, the result of people who had voluntarily become part of the transformation, learning the skills of land rejuvenation as they themselves recuperated?

John spoke of how to purify wastewater by emulating riparian areas and how to repair compacted soils with plantings that emulated succession. It was a biological and social approach to restoration, leveraging the daily activities of food cultivation, road building, water treatment and harvest, shade and shelter creation, etc. Our advice was not to install one-size-fits-all solutions, but to learn from local ecosystems how to rebuild soils, cleanse water, support biodiversity, and refresh the human spirit.

What happened next was a glimpse into how even the most jaded group can transform when they see themselves through a new lens. We pivoted and got busy covering reams of paper with designs embodying our new first principle: humans are not a plague, but a catalyst for healing. Once we allowed ourselves to imagine a new design intent—to bring broken lands and waters back to life—the answers started to flow.

There's a sense of calm that pervades John's spirit when he talks about his work, and you are about to experience it in his prose as well. I imagine him as a sea captain serene at his wheel, with eyes fixed far, far on the horizon, ably trimming his sails for each new challenge. Perhaps it's because he has ground-truthed his theory so many times: that through and *with* nature, in partnership with all six kingdoms of life, we can become agents of renewal. As he says, "We can go to bad places and do good things"; his calm belies a gigantic vision that jumps scales to encompass healing "not just of individual landscapes, but possibly the whole world."

To jump scales with ease, one must truly know the nestedness of systems in the natural world. The same rules that govern small interactions in cells and between cells have analogues in and between organisms, and repeat again when you get to the ecosystem, biome, and biospheric levels. Healing in one can trigger healing in another—from microcosm, to mesocosm, to macrocosm. John knows that when cells of living ecologies are linked, they find new ways to solve problems—as in the example he gives of cleaning tainted water, together. When you link these to larger ecosystems, network rules pertain. That is why he can move so easily from healing a pond in Massachusetts, to a citywide canal network in South Africa, to hundreds of scarred mountaintops in Appalachia, to the Sinai desert itself. Healing of water, soil, climate, and community can spread, and humans can be a nucleus for that healing.

And now it's your turn. John's manifest belief in the recuperative powers of nature, and in the ecological niche of human as helper, is something that will change what you think is possible. It will make you want to find your own wild mentors and ask what stories they might know about living on

Earth over the long haul. You'll begin to realize that the healing in *Healing Earth* is both an adjective and a verb. Through his tale of becoming nature's apprentice, John has left breadcrumbs for us to follow in the pursuit of the verb form, leading us outside and within so that we might "re-member" our own precious piece of the whole.

JANINE BENYUS
Stevensville, Montana
Summer, 2018

1

A Vision of Hope

I am writing this book based on the belief that humanity will soon become involved in a deep and abiding worldwide partnership with nature. Millions of us will commit ourselves to reversing the long legacy of environmental degradation that threatens to destabilize the climate as well as the great ecologies that sustain life on Earth. We must develop a vast stewardship initiative, which will become the great work of our time. Fortunately, there are as many ways to serve the Earth as there are people willing to engage in this vast restoration project. It includes nothing less than stabilizing the planet's climate as well as saving ourselves.

I haven't always felt hopeful. As a child, I watched with sadness as nearby woods were cut down for houses, marshes transformed into parking lots, and streams were buried in rubble before my eyes. As I became more observant I began to feel myself to be a companion to the landscape. When I saw eroded hillsides, I could feel the pain of their scars. I was saddened by the loss of animal species that I knew. The last whip-poor-will that perched on the roof of my house thirty years ago, which I never heard from again, still haunts my nights.

I had an eclectic education. At McGill University in Montreal I studied agriculture, where I was also taught evolution and ecology. I felt then that there lay great mysteries in these fields that would shape my future work and thought. My dream then of becoming a farmer would take on dimensions I could not foresee. I subsequently took a graduate science degree at McGill in parasitology and tropical medicine. My major professor, Dr. Marshal Laird, was an ecologist who studied malaria and did brilliant work in the field, mostly in the South Pacific. He inspired in me a love for working in the wild.

After graduating, I worked as an environmental consultant for a year, then went to the University of Michigan to study aquatic science and animal behavior. I was intrigued by how animals communicate underwater. I was privileged to do my doctorate under Dr. John Bardach, whose scientific curiosity had few limits. He expected as much from his students and colleagues. Under his tutelage, I made several discoveries, including how fish navigate through their sense of taste. I later uncovered their complex social communication by chemical signals. My study animals were catfishes, which had highly developed taste buds on their bodies. Their social behavior was organized by chemical communication and mediated through an acute sense of smell. This work was eventually published in the journal *Science*.

At this stage in my life, I was becoming aware of the unfolding environmental crises of our time. With my wife, Nancy Jack Todd, I would discuss the implications of such writings as Rachel Carson's cry of alarm in *Silent Spring*. The big question on our minds was, what were humans doing to the Earth and how should we respond? We had children and we wanted to ensure that they and their children would have healthy lives.

My initial response was to study the impact of toxic agricultural chemicals on the lives of marine fishes. At San Diego State University and then at the Woods Hole Oceanographic Institution on Cape Cod, I designed simulated marine environments inside greenhouses. I placed in them fishes that had observable social lives, including caring for and protecting their young. I was not prepared for what I found. At extremely minute sub-lethal concentrations of the organochlorine pesticide, DDT, parent fish would lose their ability to recognize their young. Instead of caring for their offspring, they would eat them. I also discovered other behavioral anomalies that further doomed these fish communities. Other pesticides had the same effect. This work was shattering for me. I could imagine spending my life chronicling the end of nature and did not like it, even while knowing such knowledge could become a wakeup call for some.

I wanted to design and build healthy systems that would both feed people and enhance the environment. I was determined to develop and use renewable energy sources with ecological technologies to carry on the work of society. My desire was to create positive ways of caring for people and the land. With my wife and our close friend Bill McLarney, an authority on the cultivation of aquatic foods, we formed an organization we called The New Alchemy Institute in 1969, which remained active until the early 1990s. Our motto was "To Restore the Lands, Protect the Seas and Inform the Earth's Stewards." Much to the amusement of our friends, the three of us had big dreams.

We established our institute on a thirteen-acre former dairy farm on Cape Cod. Within a few years, the property was the site of windmills, fish ponds, solar greenhouses, bio-shelters or "arks" as they were called, research gardens, agroforestry plots, and a small vineyard. It was part research center and part farm, as well as a place to envision future possibilities. It quickly became a hopeful destination for more than ten thousand visitors a year. The staff was small, talented, and very dedicated. For over a decade it symbolized the future and subsequently spawned numerous other projects and organizations around the world. The ideas had begun to take root. The story of New Alchemy and its legacy has been told by Nancy Jack Todd in her wonderful book *A Safe and Sustainable World: The Promise of Ecological Design*.

After working with coastal fishing communities for several years in South and Central America and inventing living technologies primarily in North America, I returned to teaching at the University of Vermont. I found the environmental and natural resource students bright and well educated, but their focus had been on the ills of the world and its ecological crises. To me, most of them seemed paralyzed by such knowledge and, to a certain degree, depressed. I was determined to try and change their minds about the future. One of the mottos of my course on living technologies and ecological design was "that which has been damaged can be healed" and another was "do good things in bad places." I tried to show the students that the field of applied ecology could solve problems, was practical, and in theory at least, could save the world. Their response was incredible and some of their work in class and the studio bordered on the professional. They were creative and turned on. There was such joy in the classroom.

In Woods Hole, I became friends with the brilliant biologist Lynn Margulis. She was pioneering a new dimension in evolutionary theory. It was the idea that different organisms co-evolved together and in association became something completely new. Her book, *Symbiosis in Cell Evolution*, transformed my thinking about life in concert. Her symbiotic evolution theory convinced me that I needed to develop a coevolutionary theory of design and that it needed to be rooted in ecology. Ecological systems are the extant legacy of several billion years of life's developing relationships. I wanted to decode nature's operating instructions, or at least attempt to do so.

Another person who influenced me was Dr. Margulis's scientific partner, the English atmospheric chemist James Lovelock, whom I also came to know. He and Lynn developed the Gaia theory, which declared the Earth was a coherent living being, capable of self-regulating and self-organizing on a global scale. This led to their choice of the name Gaia, after the Greek Goddess of Earth. Lovelock's book *Gaia: A New Look at Life on Earth* radically changed how many people see the world, its complexities and challenges. Also, Dr. Lovelock discussed how Earth responds as a complex living system when confronted by the destructive behavior of people who collectively cause harm on a planetary scale.

James Lovelock went on to do humanity a great service with his book *Healing Gaia: Practical Medicine for the Planet*. It is an amazing chronicle of the Earth, its history and evolution, strengths and ills, all from a medical perspective. He compares human to planetary medicine. He discusses how problems are diagnosed, which instruments to use, and what methods can correct challenges like global warming. *Healing Gaia* is a narrative of the Earth through the lens of the systems that make it healthy, including geology, soils, vegetation, freshwaters, the oceans, and the atmosphere.

This book, *Healing Earth,* looks through my eyes at my career creating recipes for healing the Earth, and hopefully it will inspire you to create your own practical responses. I hope that the reader will come to see that such healing has many pathways. It can operate at many different scales—from that of an urban window box growing leafy vegetables to vast regions being re-greened, such as the Sinai Desert or the Loess Plateau in China. Many of the solutions are based upon a variety of locations and settings. They range from cleaning up small polluted ponds to treating sewage and other wastes ecologically. They include growing diverse foods. There is often an economic dimension, as I've described in chapter 15, entitled "Appalachian Spring."

What differentiates restoration practice is the point of view of the practitioner. Take the case of carbon farmers, who use farming methods to capture and hold carbon in the ground and in vegetation. Carbon farmers are economic farmers in the conventional sense. Also they are committed to using methods that take carbon dioxide from the atmosphere and sequester it in the soils as stable carbon. In this way, they are not only feeding people and livestock, but they are also helping to stabilize climate. Such solutions are completely local, but also global in their aggregate impact upon the health of the planet. There are literally thousands of ways that stewardship can help heal and transform our wonderful Earth.

2

The Birth of an Ecological Technology

In the 1980s I was confronted with what seemed like an impossible challenge. Two of my friends had prematurely died of cancer, and I began to wonder if their illnesses could have been caused by carcinogens in the environment. I was determined to find out more, and what, if anything, could be done. One day I visited a landfill in the small town of Harwich on Cape Cod. In the center of the town dump was a series of lagoons that were filled with a fetid waste that had been pumped from septic tanks into tank trucks and brought to the site for discharge. I later learned the liquid came from household septic tanks or cesspools and from a wide variety of sources including small businesses such as gas stations, machine shops, stores, veterinary clinics, assisted living complexes, and even medical facilities. The list of sources was long.

The lagoons themselves were dug into Cape Cod's coarse sand and left unlined. The waste liquid, or leachate as it is called, would percolate down into the ground, leaving the solids behind. The solids were buried later when the lagoons were filled back in. I subsequently learned that the lagoons contained most of the United States Environmental Protection Agency's top fifteen priority pollutants, mostly carcinogens or suspected carcinogens.

Unlined lagoons for septic tank wastes

The travesty did not end there. The lagoons were dug in coarse sand that was very porous to the liquid migrating through them. The problem was compounded by the fact that they were situated directly above the groundwater that was used by the town for drinking. I was appalled by the potential of the lagoons to contaminate the groundwater and subsequently a thirteen-acre pond and a stream below the landfill. The practice of holding wastes in unlined ponds has now been stopped, but for most of the twentieth century such insidious contamination of the groundwater went on unabated throughout the United States and around the world. This is not an isolated story.

I began to inquire with wastewater treatment professionals why these wastes were not treated and learned that they are typically 40 to 100 percent more concentrated than sewage. As a consequence, waste treatment plant operators do not normally want to handle them because they are so toxic and can interfere with sewage plant operations. Further, there did not seem to be a cost-effective technology on the market to treat such waste and most communities could not afford much in the way of treatment. There was a technological gap that needed to be filled somehow, so I decided to design a solution.

The question became one of where I should turn for inspiration to develop such a technology. As an ecologist, my bias was, and is, that nature needs to inspire design. There is a very pragmatic reason for this. Over the past three plus billion years, nature through trial, error, and adaptation, has experimented with an unimaginable diversity of problems and stresses. It "knows" how to cope with most of the toxins in the environment by using complex systems composed of members of all the kingdoms of life to solve its problems. Life has evolved this way.

On Earth, the sun sustains life. Sunlight is the primary energy source for almost all of the ecosystems on the planet. I decided that our natural systems technology should be based upon solar power. Mine was a clear break with tradition, as most conventional waste treatment technologies do not use photosynthesis in the purification of contaminated water.

The design was based on cylindrical tanks made from a thin fiberglass material that allowed sunlight to penetrate the sides as well as the surface. This created a three-dimensional solar effect and the walls of the tanks supported vibrant algae-based communities. These tanks were connected in

Solar aquatic river for waste treatment

series to create a facsimile of a flowing "river." The facility was built at the Harwich landfill.

My next pivotal decision was to introduce a great diversity of life forms into the solar algae tanks. My reasoning was that only through introducing thousands of different species of organisms would we find the right ones for creating biological communities capable of dealing with all the toxic compounds in the wastes. To achieve this end, my colleagues and I gathered living organisms from over a dozen ecosystems. They included local salt marshes, streams, ponds, vernal ponds, wet spots in the woods, and even a pig wallow on a local farm. We introduced the organisms into the tanks. The water within was then recycled to distribute the diversity of species throughout the whole system before adding the waste.

Once the waste was added, the system very quickly self-organized and self-designed itself into completely new ecologies that were specifically adapted to its contents. The system went even further. It created unique ecologies for each stage in the transformation of the waste stream. Each tank in the series was slightly—and sometimes very—different from the one before it.

Biodiverse communities in the tanks

Lynn Margulis and her students at the Marine Biological Laboratory in Woods Hole studied the communities that developed in the tanks. She recognized the various life forms within the tanks but, to her great surprise, the communities that had formed on the walls of the clear-sided tanks were completely new and unique. The possibility that this kind of ecological invention could actually happen had been predicted in 1972 by the ecologist H. T. Odum in his visionary book *Environment, Power, and Society.*

More by accident than design I had inadvertently included representative species of all the kingdoms of life in the systems. It would be years before I began to appreciate the significance of this strategy. I would also learn that a diversity of organisms from a variety of parent ecosystems could produce systems with a meta-intelligence that had a highly specific ability to self-organize, self-design, and self-replicate. They were capable, in fact, of surviving through long periods of times, possibly centuries, with minimal human support.

I also started to design analogs of different parent ecosystems directly into the technologies themselves so that they have equivalents of a marsh, pond, and stream stages that were interconnected. The combination of organisms from different parent ecosystems and the analogs of the ecosystems within the technology itself produced the meta-intelligence mentioned above. This is profoundly important because it carries within it a legacy of vast reaches of evolutionary time. I was seeing the potential for totally new kinds of technologies such as "eco-machines" in which humans were the junior partners in the endeavor. Eco-machines are living technologies that are engineered to undertake a variety of tasks from growing foods to treating wastes and repairing damaged environments. Ecological design was destined to become a new kind of design science. That said, we still have much to learn about how these eco-mimetic or living systems work as they do.

Our test and demonstration technology at the landfill performed incredibly well. The waste traveled through the system in ten to eleven days. By the time it reached the end, it was crystal clear. The quality of the water was very good and met drinking water standards for being free from heavy metals. We subsequently learned where the metals were stored. The bulk of their mass found their way into the algae-dominated communities on the walls of the tanks near the upstream end of the process. The priority pollutants or toxins in the system were removed to levels that were below detection in the water at its point of discharge. Only a trace amount of one of the chemicals, toluene, was detected and it had been removed by 99.9 percent. We ran the prototype eco-machine from spring through the summer and into the fall.

The project then took on an ironic twist. The pilot study had been funded by the Massachusetts Foundation for Excellence in Marine and Polymer Sciences. Nevertheless, a regulatory entity, the Massachusetts Department of Environmental Protection, decided to fine me thousands of dollars for carrying out the experiment. Because of the success of the project I was shocked and bewildered when my name and organization appeared on the front page of the *Boston Globe*. I was declared a scofflaw. It turned out that I had made a mistake. In Massachusetts a scientist is apparently not allowed to design a pilot waste treatment facility. Only a civil engineer who has the letters "PE" (professional engineer) after his or her name can do so.

William Reilly, the administrator of the United States Environmental Protection Agency (EPA) in Washington, subsequently heard about my plight and sent one of his experts to look over the facility and the experiment. The

Clear and clean water at the end of the process

Year-round greenhouse-based system subsequently built on the site

expert declared it *bona fide*. The EPA subsequently honored me with the first Chico Mendes Memorial Award for the project and the lawsuit against me was quietly dropped.

Later a more permanent facility was built on the site. It was housed within a greenhouse so that it could be operated year-round for several years. The facility was subject to intense scrutiny. It performed well and the technology was eventually permitted by the state.

I cannot describe my joy that first summer at seeing the waste transformed into clean water. The experience gave me the confidence to explore new problems and do so widely throughout the world. With this first eco-machine we learned that it is possible to do good things in bad places. Harmful chemicals can be treated. We continue to explore the number of chemicals that living technologies can render harmless.

3

Steps to a Theory of Ecological Design

I am an optimist at heart, but I have tested my optimism in places that have been degraded and almost destroyed by abuse from humans. Toxic waste sites, polluted waters, and ravaged environments have been home to some of my projects. My efforts have resulted in some remarkable changes. This stewardship road has turned into a path of learning and of wonder at the power of the natural world to benefit humanity, if only we could understand its instructions and work with the living world in a new kind of partnership.

Here I would like to introduce you to the first principles of environmental healing. On one of my early projects, in which I set out to clean up toxic waste water that was contaminating the groundwater of a town nearby, a scientist friend asked me, how did I know what I was doing, and where did I get the knowledge to understand the system with which I was working? I told him that I did not have the knowledge, nor did anyone else on Earth. I told him I was relying on the intelligence within the three-plus billion-year legacy of life on Earth. I hoped that diverse nature would work in concert with my efforts and accomplish the ends I desired. At best, I was being a biological choreographer and my dancers were the ecological components I brought to the performance.

I intuited that perhaps for the first time in modern history our knowledge of the natural world has reached the stage where we can look into nature and see a coherent body of ecological information there. Life's inner workings are being decoded and its extraordinary complexity revealed. We are learning that nature has a set of operating instructions of immense significance, which I believe are critical to humanity's future.

We know enough at this juncture to ask two fundamental questions: can we use contemporary ecological knowledge, combined with a great diversity of life, to create the technologies of the future, technologies that are powered by sunshine, recycle and reuse energies, that don't produce pollution as a byproduct, and where external inputs of scarce materials are reduced to a minimum? Secondly, is it possible for us to employ these technologies and techniques to take on the task of sustaining society, including its food, shelter, and energy needs?

To answer these questions, I propose that a new theory of design, based upon ecological principles, is needed. Such an ecological design theory requires blueprints from the natural world, a deep understanding of the significance of complexity, and the ability to recombine species from all the kingdoms of life into new associations or assemblages that can carry out a variety of tasks designated by ecological engineers. As I mentioned before, in 1972 the insightful Howard T. Odum[1] predicted that an ecological design science was in fact possible, and that it could provide the basis for a technological revolution suited to an age of resource limitations and a degraded planet.

In the natural world, systems are made up of a network of ecosystem traits or behaviors that may well be universal. The simplest way to understand these traits is to observe their workings visually by creating a microcosm from scratch. For example, if one filled a five-gallon glass jar with several buckets of water from a pond and placed the jar in the sun, a remarkable unveiling would happen. The pond water will contain hundreds of species of life forms from most, if not all, of the six kingdoms of life, Eubacteria, Archaebacteria, Protista, Plantae, Animalia, and Fungi.

At first, chaos and lack of order will characterize life in the bottle. Then sediments in suspension will slowly descend to the bottom to form a substrate that will host a community of organisms that depend upon sediments. As the days go by, different life forms will begin to make themselves known: algae blooms occur first and turn the bottle green; then tiny pinhead-sized animals will emerge to graze the algae. Within a week, minute shoots of aquatic plants will poke up through the sediments. Insects and aquatic mites swim in the water column. Snails begin to show up and make tracks as they graze the walls. Round worms and leaches will make themselves known.

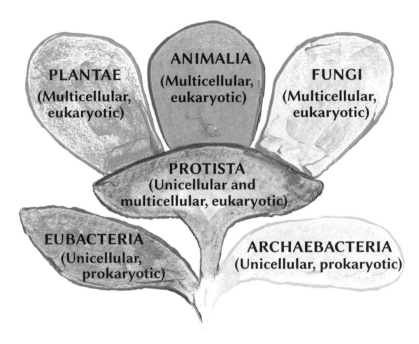

The kingdoms of life

Mosses and fungi appear. Later tiny fishes will grace the water and a freshwater sponge is observed growing on a stick in the jar. These organisms, at one stage or other of their life cycles, were in the original seeding of the bottle. However, their roles in the new system will emerge over time.

There are many variations of this aquatic theme, depending on the time of year, the source body of water, and the weather under which the seeding takes place. The newly created ecology in the bottle is different from any wild ecology, but it shares species with the wild progenitors. The ecology that exists in the sunshine-bathed glass bottle is a powerful expression of nature's ability to self-organize, self-design, and self-repair. Over time the microcosm of life in the bottle will develop the ability to self-replicate, and some of the species, but not all, will adapt to the conditions within their environment and persist long into the future, perhaps for decades.

We are beginning to learn that these forces found in the contained microcosm can be utilized to great effect in the design of living technologies, and they lead to what I have called the First Order of Ecological Design. So far, we have identified three orders of ecological design, each with its own unique set of attributes. However, the First Order of ecological design provides the building blocks for the Second Order and both provide the framework for the Third Order that encompasses spans of time, economics, and institutional structures.

First Order Ecological Design is to a large degree a technological domain, a creator of living technologies. Using the "intelligence" of several parent ecosystems combined with a set of design principles, an ecological designer can devise a living technology to serve a specific purpose or purposes. These technologies can be designed to undertake tasks such as generating fuels, growing foods, treating and recycling wastes, detoxifying dangerous organic compounds, or repairing damaged ecosystems.

I have found through years of trial and error that to optimize their performance a few rules of thumb should be followed: first, at least three parent ecological types should be incorporated into the design. For example, an ecological technology might have marsh, stream, and pond analogs within it. An eco-machine (a type of living technology) that treats sewage will have, at the beginning of the process, a dark ecology, one that is without light and starved of oxygen. Liquid from this fermentation-driven system will subsequently flow into a second ecology derived from marshes, that are in turn followed by a pond- or lake-derived system specifically modified with technological support, such as supplemental aeration, for the management and removal of nutrients, contaminants, and pathogens from the waste water. The goal for the eco-machine is to produce clean water suitable for reuse or discharge into a natural environment. The best systems can have an economic status incorporating valuable byproducts into the process, including ornamental fishes, cut flowers, and a variety of trees and shrubs.

The design of an eco-machine to produce human and animal foods through the recycling of organic materials that are often considered wastes, such as spent grains from a brewery, employs the same rules of thumb and initially the same diversity of organisms. However, the food webs within the eco-machine will be different and will often serve different functions in the newly emergent ecosystems.

Essential to the optimal functioning of living technologies, such as an eco-machine, is a set of design and operational principles that were developed by us, and others, through years of system performance optimization. Here I am going to list the thirteen guiding principles with a comment on their role or function in the design of living technologies. There may be more that have yet to be discovered, but these are sufficient to understand the workings of these systems. For a more complete explanation, see Todd and Josephson (1996).[2]

1: MINERAL DIVERSITY

Years ago, I was involved in a modest reforestation project in Costa Rica on the Pacific slope of Guanacaste province. We were planting trees in hard, overgrazed ground. Half of the seedlings were given some compost in their planting holes and the other half received the compost plus a cupful of ground-up volcanic rock from a nearby quarry. The young trees with the rock powder grew so much better in the early stages of our experiment that the lady carrying out the planting and care of the trees, without warning us, stopped the compost-only tree plantings and gave all the trees the compost plus rock treatment. While I understood her concern for the seedlings, she had taken away the original foundation for our experiment. What we did learn was that degraded soils benefit from rock powders.

Ground-up rock powders from different parent rocks such as granite, sandstone, volcanic materials, and so forth provide a template for microbial "invention" and diversification. Rock mineral powders have been shown to be critical in the reestablishment of degraded soils and landscapes and in optimizing agricultural production in certain areas.[3]

2. NUTRIENT RESERVOIRS

Nutrient surpluses in a system provide a steady, slow-acting reservoir essential to sustain complex systems. Ecosystems naturally develop their own reservoirs over time, such as the accumulation of rich soils or bottom sediments in water bodies. Surplus nutrients in a slow-acting form provide optimal stability for an eco-machine. In our systems, rich organic sediments are the norm.

3. STEEP GRADIENTS

I have found that rapid and steep transitions within living technologies improve their performance. These are somewhat difficult to describe, but a waste stream flowing quickly from an acid environment to an alkaline one through a low-oxygen to a high-oxygen zone would be examples of steep gradients in a system. Abruptness, or dramatically different transition zones, is the key here.

4. HIGH EXCHANGE RATES

High exchange rates are also not easy to describe, but the objective of exchange rate engineering is to maximize contact, for instance of wastewater, with active biological surfaces. I have designed ecological fluidized beds that circulate large volumes of water through semi-buoyant media. These technologies support higher plants on their surface and the plant roots penetrate the media. High exchange rates allow the designer to shrink the footprint or space of their technologies.

5. RANDOM AND PERIODIC PULSES

Naturalists have observed a strange phenomenon in some desert areas. Two identical-looking areas that have similar amounts of rain but different rainfall patterns evolve differently. Those areas that have erratic and unpredictable rainfall patterns contain and support many more species than those with highly predictable rains. Randomness influences the fate of a system.

Ecosystems have been shown to be more robust when they exist in unpredictable environments. Artificially introduced random and periodic pulses, such as erratic light regimes or flow rates tend to make living technologies more adaptive. Sadly, very little research on this subject has been carried out.

6. CELLULAR DESIGN

Cells in nature are the building blocks of life. They are unique insofar as they integrate two opposing tendencies to the benefit of both. A cell is whole and complete. It has all the basic functions of life within it. A cell can feed, excrete, and divide into new cells. It is independent to a degree. But it is also interdependent with the cells around it, outward to the organ within which it resides, and then to the organism itself. The genius of nature is found in its design of autonomous building blocks placed in highly coupled interdependent settings. Engineers rarely design this way, but ecological engineering may well be predicated upon it. Hence any eco-machine, or similar technology, will be made up of cellular units that together comprise the whole system.

7. SUBSYSTEMS FROM DIFFERENT PARENT ECOSYSTEMS

I have discussed parent ecosystems previously under basic rules of thumb. If, as an example, an eco-machine has wild parentage from three ecosystems such as a pond, a stream, and a marsh, it will contain different components or subsystems that are, in essence, analogs of the parent ecologies. Each component will house many of the organisms and design elements characteristic of the wild type. The pond will have open water over a bottom sediment community, and the stream component will be dominated by flowing water and organisms associated with movement. For its part, the marsh component will be dominated by emergent plants. Oxygen transfer from the higher plant roots, low-oxygen-tolerant microorganisms and burrowing creatures will drive the biochemical reactions characteristic of mucks and living organisms within the system.

When different parent ecosystems are combined in the design of a living technology, a robustness develops. Combined they seem to create a "meta-intelligence" with capabilities that would not be predicted from the subsystems alone. The living technology begins to behave in ways that are greater than the sum of its parts. Their ability to withstand perturbations and toxic onslaughts is greatly increased and their performance rates go up. "Meta-intelligence" in ecologically engineered systems should become a fertile field of research and investigation.

8. BACTERIAL COMMUNITIES: ECOLOGICAL WORKHORSES

It took me many years to truly appreciate the pivotal role of microorganisms in healing the planet. Until I met my mentor Lynn Margulis, the great evolutionary biologist, I did not understand the full significance of bacteria and other microscopic organisms. Four of the six kingdoms of life comprise the microbial realm that is the cornerstone of life on Earth. The two bacterial kingdoms, the Eubacteria and the Archaebacteria, are made up of an enormous diversity of organisms that are unicellular and do not have a differentiated nucleus. Archaebacteria were traditionally seen as organisms of extreme environments, but are now known to also be extremely common elsewhere, particularly in the ocean. The best known of this group are the methanogens, the organisms responsible for biogas production and the subject of much current energy-related research.

Eubacteria are widely distributed and common and include the heterotrophs, bacteria that absorb organic materials; the chemotrophs that break down inorganic matter, such as rock powders; and the autotrophs that make their food through photosynthesis. Because of their color the autotrophs are sometimes called blue-green algae or cyanobacteria.

It is truly difficult to comprehend the significance of the two bacterial kingdoms in the workings of the planet and it is almost as difficult to understand comprehensively their roles within ecosystems. They have been on Earth more than 3 billion years and are the true ecological pioneers. Their diversity of species is enormous and as new identification methods are developed their species numbers grow. Some authors suggest there may exist over a million species. From my perspective, bacterial diversity is essential to the functioning of an eco-machine. The idea that there are magic-bullet bacterial species that can do our work for us is beginning to die out at last. Life works in concert. The vast majority of bacteria cannot be easily cultivated in the laboratory. It takes an ecosystem to support them.

Photosynthesis that uses carbon dioxide as a fuel and produces oxygen as a byproduct is a major regulator of planetary dynamics. The same regulatory functions hold true for living technologies including eco-machines. Bacteria encompass some species that undergo photosynthesis, but the process really comes into its own amongst the kingdom Protista.

9: KINGDOM PROTISTA

The Protista mostly are microbial organisms. It is a huge and diverse kingdom characterized by the fact that all of its members have a true cellular nucleus.

They can be unicellular, colonial, or multicellular. The Protista is composed of six major groups of organisms. Most, but not all, are microscopic.

PROTOZOA: The first group is the Protozoa, animal-like and mobile. The amoeba is a member of this group and perhaps its most infamous species is *Plasmodium vivax,* the parasite that causes malaria. However, a number of species have chloroplasts and photosynthesize. They are very important in the regulation of microbial communities and in the reduction of bacterial pathogens in wastewater.

DINOFLAGELLATA AND EUGLENOPHYTA: Secondly there is the algae group that includes the Dinoflagellata, most infamous because they cause red tides, turning seawater red with their vast numbers. Also in this group are the Euglenophyta with some members capable of propelling themselves towards light to photosynthesize more actively.

CHLOROPHYTA: From the perspective of an eco-machine the next group, the Chlorophyta or green algae are very important, as their diverse species are found in moist environments. They play a multiplicity of roles in most, if not all, of the living technologies.

PHAEOPHYTA: Next are the Phaeophyta (brown algae) and Rhodophyta (red algae). Both are important members of photosynthetic community. The brown algae that are known as kelp have recently been shown to be important regulators in the global carbon cycle.

MYXOMYCETES AND OOMYCETES: The final group in the Protista is composed of fungus-like organisms, known unflatteringly as slime molds and water molds. At one time, they were classified with the fungi, but now the distinctness of this group of organisms is recognized. They are mobile and absorb nutrients and play an important role in metabolic turnover in many environments. They do not photosynthesize. I suspect that they will prove to be important internal regulators in many emerging technological systems.

At this point in time our knowledge of the various functions of the Protista in the dynamics of complex living systems is, in relative terms, quite scant.

While their importance is clear, their use in technological systems has been scarcely explored.

10. HIGHER PLANTS: KINGDOM PLANTAE

The higher plants are included in another kingdom of life, the Plantae. They are of immense photosynthetic importance. The plants evolved from the Protista about 500 million years ago. They represent all the land plants including mosses, ferns, conifers, and flowering plants. They are characterized by three attributes: first their cell walls contain cellulose; second, they are composed of delineated structures (organs, roots, stems, and leaves); and third, they contain chlorophyll a and chlorophyll b.

Several decades ago the importance of higher plants in the design of living technologies was only appreciated by a small group of biologists and wastewater engineers working with constructed wetlands to treat various waste streams. But as our knowledge of the significance of higher plants in hosting microbial communities and in the formation of soils and in the sequestering

Interior of the South Burlington eco-machine for sewage treatment

of carbon has grown, this old bias against higher plants in technological settings has begun to wane. We have been able to evaluate hundreds of species of plants for their ability to support wastewater treatment. Part of the evaluation included pest and disease resistance, growth and flowering, root depth and density, seasonality in contained solar environments, cold/heat sensitivity, and their secondary economic value as commercial byproducts.[4, 5]

11. FUNGI-DECOMPOSITION, NUTRIENT CYCLING, AND EXCHANGE

The Fungi evolved from the Protists about 1 billion years ago. The number of species described to date is currently about 100,000, but several estimates put the total number of species at 1.5 million; a recent study suggests this number may go as high as 5 million.[6]

This truly unfathomable diversity is important to us for a multitude of reasons; first the number of biochemical pathways they can produce simply beggars the imagination. Enzymes produced by fungi represent a major line of defense in breaking down complex molecules, hence their emerging pivotal role in the field of environmental detoxification and restoration. Second, the vast diversity of species suggests that there may be an almost unlimited number of future tasks for them in the ecological designer's toolbox. Paul Stamets explores this exciting frontier in his book *Mycelium Running*, which has taught us the significance of the importance of fungi in ecological engineering.[7]

The fungi are characterized by having a true nucleus (eukaryotic) and nonvascular organisms that reproduce by spores, which are most often wind-disseminated. They produce both sexual and asexual spores depending upon the species and environmental conditions. Their cell walls contain chitin and not the cellulose found in plants. Typically, they are not mobile, although a few have a motile phase.

Most of us know the fungi through mushrooms, some of which are extremely tasty while others can be deadly. They are also well known for their importance in medicine (antibiotics) and are becoming increasingly appreciated for their role in human health and for the prevention of diseases including cancers. We also appreciate their significance as single-celled yeasts in the fermentation of wine, beer, and soy sauce, among many other delicious foods. Fungi cause diseases in plants and animals, ourselves included, but they also play a positive role in the protection of plants and animals. It is not that nature proceeds by ironies, but that these contradictions are the consequence of the enormous biological diversity that exists on Earth.

My colleagues and I have always had fungi within our living technologies. They were there because of our seeding the systems with life from at least half a dozen local wild ecologies. But in recent years we have begun to appreciate that specific species of mushrooms, at distinct stages of their life cycles, produce chemicals that we can use to a good end, such as degrading Bunker C crude oil. Bunker C crude, or #3 oil, is the sludge, the densest material left over from petroleum refining. It is used to power ships' engines and in earlier times was a fuel for factories and a heavy equipment lubricant. Bunker C has ended up contaminating canals, rivers, and groundwater along old industrial corridors. We devised a small eco-machine to treat this material and one of the cellular elements was dedicated to the production of enzymes from the mycelia or threadlike structures produced by a mushroom. This fungal cell component dramatically increased the rate at which the crude oil could be broken down into less toxic and more assimilative chemical compounds. The system worked. At this historical juncture, we know little of the ecological roles played by the 5 million-plus species of fungi, but we do know they are important. As more and more species are studied and applied to task-specific ends, the more we will be able to ascertain their true significance.

12. ANIMALIA—ANIMALS, THE UNKNOWN ACTORS IN THE ECOLOGICAL PLAY

In the *Biology of Wastewater Treatment* published in 1989, author N. F. Gray barely mentioned the role of animals in the purification of sewage water. In *Ecological Aspects of Used Water Treatment,* 1983, authors Curds and Hawkes scarcely discuss snails and treat insects associated with sewage in a derogatory manner, and neglect the rest of the kingdom Animalia, comprising between 9 and 10 million species. Luckily, in the past three decades many advances in understanding have allowed us to appreciate the role of animals in ecological systems.

Anybody who has investigated healthy soils or the sediments of ponds or lakes has seen firsthand the extraordinary diversity of animals that exist in these environments. They burrow, till, and excavate, moving vast amounts of materials around and through their own byproducts they capture and concentrate nutrients within complex systems.

Animals have played key roles in the functioning of eco-machines, and new roles are being found for them all the time. For example, pond snails of the families Physidae and Lymnaeidae are excellent grazers of attached algae and slime in many different types of eco-machines. They are constantly cleaning various surfaces, including tank walls, and in some systems their grazing permits light to pass into the interior of translucent tanks, thereby increasing photosynthesis within.

The snails play another role. Whenever toxic materials get into a living technology, the snails leave the waste stream and climb up onto the underside of the leaves of higher plants just above the water surface. They function as an early warning system, allowing an operator to respond to the toxic threat in a variety of useful ways.

Many animals are filter feeders, including small fish characterized by finely toothed gill rakers in their mouth cavity. This allows these animals to filter free-floating microscopic materials directly out of the water and in so doing speed up the purification process. Tiny shrimp-like zooplankton, including members of the genus *Daphnia,* do the same thing as they swim throughout the water column. Bivalve mollusks, such as oysters, clams, and mussels, are superb at removing all kinds of microscopic materials from seawater and their freshwater cousins have the same remarkable qualities. We have used to good effect native freshwater clams of the genera *Unio* and *Anodonta* in eco-machines designed to help clean up lakes.

Most aquaculture facilities and waste treatment plants produce copious amounts of sludge and noxious materials made up of dead and dying bacteria and their associated communities. Sludge removal and treatment is an expensive chemical and energy-intensive process. However, these same organic materials represent a superb food source for certain kinds of fish including the valuable Japanese Koi, a greatly esteemed ornamental fish. We have found that the Koi, a member of the common carp family, can utilize sludge in their diet and effectively carry out sludge management in ecologically engineered sewage treatment plants.

The roles of animals in living technologies have scarcely been investigated and their potential as allies in living technologies can only be surmised. However, we know enough to say that their importance will grow significantly in the future. What is needed now are studies of the natural histories of potential animal candidates for living technologies. What is also needed is a historical review of the natural history literature especially from nineteenth- and twentieth-century archives that predate modern electronic scientific information storage and communication. This material represents a gold mine of "lost" information.

13. MICROCOSM, MESOCOSM, AND MACROCOSM RELATIONSHIPS

A number of philosophical writings, including those from the Hermetic tradition, suggest that there exists a continuity from the smallest realms discovered by physics and biology outward and beyond Earth to encompass the whole known universe. They point to a nesting of realities beginning with the invisible dimensions of microscopic systems mirrored within the larger workings of nature. The microcosm is the smallest scale employed by ecological engineers. Microcosms have been used to develop models of the complex systems we call ecosystems. They have also been used to study the fate of ecosystems that are, or might be, exposed to changes in climate or toxic stresses such as air pollution. They can be windows into the future.

A mesocosm occurs at a larger scale and encompasses more than just a single parent ecosystem. They are often designed to carry out "work" for society. Eco-machines are examples of mesocosms in an applied setting.

The macrocosm is a larger-scale phenomenon; examples might be a forest, a prairie, or a lake. Macrocosms have a time dimension that is more pronounced than in microcosms and mesocosms. This time dimension is often expressed as succession in ecosystems, the processes whereby a bare field becomes a meadow and then over the years, if there is enough rain, the meadow becomes a shrub-dominated landscape and finally after decades the whole system transforms itself into a forest. The time dimension in this example of succession involves an "architectural" shift on the landscape as the grasses and flowers are replaced by trees and forest organisms.

What is important for the ecological designer is that each of the time-dependent stages is potentially very useful in human and economic terms. It is possible to create on the landscape a succession of human and economic activities that mirror the ecological ones. At this stage, ecological or eco-mimetic design can be directed into new economic and social territory. In this setting, living technologies, including eco-machines, become part of a much larger and more complex ecological system whose functions radiate outward to encompass landscapes and human economic activity with their own succession built in.

In this scenario, the living technologies are examples of First Order Ecological Design. The next stage, Second Order Design, brings together processes and materials that are not normally connected. Industrial ecology as a discipline falls into the Second Order category. So do agricultural eco-parks, most of which are being developed in urban settings.

However, Third Order ecological design attempts to ecologically fuse First and Second Order Design into a much broader frame of reference that includes economics and social institutions. It is my belief that durable and sustainable economies, in an age of resource limits and information richness, can replace the extractive and environmentally destructive technologies and infrastructures of today. We are beginning to envision and initiate the large-scale transformation and restoration of whole regions that have been ecologically crippled by centuries and even millennia of misuse and abuse. I am convinced that an ecological renaissance is emerging in our time. Its promise is climate stabilization and a living bounty for us all. I return to an elaboration of Third Order Ecological Design in chapter 15.

4

The Edge of the Sea

On a summer day, one of my favorite things is to row my skiff out of Great Harbor in Woods Hole on Cape Cod and head to a small island called Devil's Foot. It borders the passage between Nantucket Sound to the east and Buzzards Bay to the west. The tides roar through, in excess of five knots, and both fish and fishermen tend to congregate there. For many of the locals, however, the greatest excitement of the place is trying to predict which of the yachts passing through the channel will get swept onto the rocks when they misjudge the current velocity and direction.

In the shallow tide-swept waters surrounding Devil's Foot are waving seagrass meadows. They are composed of a marine plant known as eelgrass, *Zostera marina*. Eelgrass forms very dense stands and seems to blanket the bottom. Most swimmers and bathers tend to avoid eelgrass meadows as the ground is often soft and mucky and it can be difficult to swim through the leaves of the plants. But snorkeling is another matter. I remember floating over eelgrass and peering down into their interior for the first time. It was if a whole new realm were opening before me. I was struck by all the life forms just inches from my eyes. The diversity of life was beyond anything I thought imaginable along New England's shores. I quickly recognized the pipefish and the seahorse, which are both unusual fish in that the males nurture their young in pouches on their bellies. There were also bay scallops, blue crabs, snails, barnacles, and down in the muck, hard- and soft-shell clams. There was also a variety of small crabs including the comic-looking hermit crabs, which are beloved by children wading along the shore. As time went by I noticed more exotic organisms such as tubeworms, bryozoans, sea squirts, and sponges. I also saw young fish including flatfishes and other species that I could not identify but only admire.

When I snorkeled away from the eelgrass beds and over to sandy or pebble-bottom areas, it felt as though I were swimming over the moon. Compared to the seagrass meadows they seemed devoid of life. There was something going on in the eelgrass communities that I wanted to decipher and decode. Little did I realize that I was looking at a community that was powerful and at the same time very vulnerable and one of nature's penultimate ecological engineers. This community was going to become one of my most influential instructors in design.

There are five families of seagrasses that are descendants of terrestrial plants that recolonized the ocean somewhere between 65 and 100 million years ago. They are not true grasses from the family Poaceae, but are closely related to the lilies or Magnolyophta. The eelgrasses are members of the family Zosteraceae.

Seagrasses are found throughout the world. In terms of being beneficial to humanity, they rank among the highest of all the ecosystems on Earth. Seagrass meadows store about 15 percent of the ocean's total carbon. On a per acre basis they consume twice as much carbon dioxide as rain forests. Each year seagrasses sequester about 27.4 million tons of carbon dioxide.

Eelgrass, like all seagrasses, are ecosystem builders. The leaves are hair-like and can reach up to three feet in length. They grow in extremely dense associations and spread by rhizomes. At the nodes are clusters of roots that anchor the plant to the substrate.

The eelgrasses' leaves slow the movement of the seawater and catch debris, which settles out to be transformed by resident biota into a rich substrate. This substrate in turn becomes the basis of diverse and productive food webs. During the day the plants produce oxygen in abundance and have been called the lungs of inshore marine waters.

Seagrass leaves harbor a rich epiphytic or attached community that includes algae, sessile invertebrates, snails, and shrimp-like amphipods. Because of the complex and dense structure of their leaves, they serve double duty as a refuge from predators for many resident organisms. Shrimp are one of the benefactors of this eelgrass trait.

The plants directly and indirectly further host a great diversity of organisms. There are filter feeders that help manage water quality and include clams, mussels, and barnacles, which remove plankton and other material from the water, as do sponges, bryozoans, and hydroids. There are also the members of the community that feed upon the detritus and debris that accumulates all the time on the bottom. These include crabs, worms, sea cucumbers, and sand dollars. My favorite seafood, bay scallops, play an important role here. Their diet is detritus-based. The young scallops live on the leaves of the eelgrass until they are large enough to assume a bottom-dwelling existence. Snails too graze on the algae that grows on the eelgrass leaves. Periwinkles, *Littorina,* are particularly effective attached algae-grazers.

Eelgrasses additionally provide shelter for many species of fish when they are young, including such commercial species as the Atlantic cod and striped bass. As a nursery for the fish community it is second to none. Each of the plants and the creatures that reside in the eelgrass community is an important actor in the vast ecological interaction of the sea.

Scientists have speculated for years about a symbiotic relationship between the seagrass such as *Zostera* and bivalves including the bay scallop. Their speculations were based upon the fact that when eelgrass communities die, the scallops tend to disappear. Recent discoveries have shown this symbiotic relationship to be much more complex with a three-way trilogy between the seagrass, the clams, and an interesting group of bacteria.

Normally the accumulation of organic matter in the seagrass meadows would cause toxic sediments to form, producing high sulfide levels that are poisonous to plants and animals. However, on the gills of some species of clams live sulfide-oxidizing bacteria, which enhance seagrass production and biomass. The bivalves profit in turn from the accumulating organic matter and the oxygen released by the grasses including from their root systems. This three-way symbiosis allows for the conversion of what would be a toxic wasteland into a veritable garden in the sea. It is coevolution at its most creative.

As I mentioned above, however, the eelgrass communities, which are the source of so much oceanic health and bounty, are also very vulnerable, especially to human activities. Seagrasses are in global decline. In recent decades over 12,000 square miles of sea meadows have been lost. The causes are many and most are linked to humans. Pollution is a major contributor. Storm water runoff increases the cloudiness of the water while reducing the light getting to the plants. Pollution has the same effect, but through a different ecological pathway. All the added nutrients in the water cause plankton boom-and-bust cycles that block the sunlight reaching the plants, reducing photosynthesis and depriving the communities of their oxygen. Once the eelgrasses die, the whole ecosystem collapses. What remains is a cobble bottom that hosts a meager shadow of its former community.

Under stressful conditions a slime mold, *Labrynthula zosteracea*, a wasting-disease, can cause massive die-off of eelgrass. It has been shown that eelgrass decline caused bay scallop populations to be reduced by over 90 percent. It is still not known whether the wasting-disease is related to pollution and other human-induced factors.

It is known however that fishing by dredging and trawling also kills eelgrass communities. Aquaculture and coastal development are villains as well. Some marine biologists point to overfishing as a significant cause in the decline of the meadows. Diminishing numbers of some fish species result in an increase in the growth of algae that block sunlight to the plants.

There is a powerful lesson in all of this for the ecological designer. On the one hand a healthy seagrass community and its behavior boggle the imagination with its ability to foster life. On the other its essential need to live in very clear water causes it to have a brittle nature. In normal times the entire community is responsible for the water's clarity, but when a tipping point is reached, the ecosystem topples into a runaway environment. We need to comprehend this and design based upon the self-regulating parameters of the system itself.

It occurred to me one day that eelgrass communities might be one of the ultimate models for the design of new systems for the culture of food. Across great reaches of time these communities have solved many of the same problems that ecologically inclined farmers also need to solve today. My challenge then was to begin to decode how they worked and as much as possible adapt their strategies to the design of completely new systems, which would have to support a diversity of foods with economic value. My first design in the

1970s was an engineered ecology that was a freshwater system. Several years later at the Woods Hole Oceanographic Institution I would work on saltwater analogs as well.

A FRESHWATER AQUACULTURE SYSTEM

My first eelgrass community-inspired design was for growing foods including fish and vegetables and crops of salad greens as well as herbs. It was a freshwater system. It was composed of four cells or modules linked together and connected by flowing water and the exchanging of nutrients, plankton, and purified water.

The process of design for me begins with determining how the new culture system can support a high degree of diverse life. Most aquaculture systems are the exact opposite from a design point of view. They are often highly simplified systems maintained through energy-intensive mechanical and chemical support components. The best way to achieve a diversity of organisms is through cellular or modular design, with the separate cells having distinct roles in the system. The subcomponents are connected. Flowing water is the link or binding element linking the cells to each other.

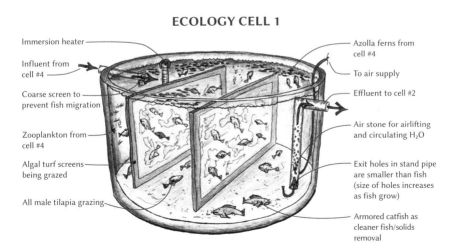

ECOLOGY CELL 1

Step two involves identifying the basic roles of the various life forms that will be needed to sustain the ecologically engineered ecology. This involves the practice of ecosystem assembly. Organisms that use sunlight, such as algae and higher plants, for example, are at the base of the food web. They provide oxygen, utilize CO_2, and are the system's primary producer of new biomass. Next are the organisms that are filter feeders such as clams. They help consume organic matter, clarify the water, and, as we have seen, in some cases detoxify it with their bacterial partners. There are also the grazers such as shrimp and snails as well as fish and then there are the detritivores or bottom-dwelling organisms such as worms and crabs and certain species of fishes. Not to be overlooked are the microorganisms including the bacteria that facilitate all the processes within the system. Finally, there are the predators, including humans, which help regulate the overall numbers. One of the best descriptions of the practice of ecosystem assembly can be found in the book *Dynamic Aquaria* by Walter Adey and Karen Loveland.[1]

Step three involves identifying the key species that will fit the various roles in the system. This process, sometimes a very intuitive one, is based

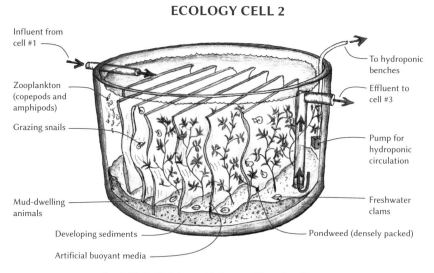

Food Chain Building with Stage 1 Water Purification

upon personal knowledge of the natural history of various organisms. It behooves one to pay close attention to the life stories of organisms including the fungi, plants, animals, and even the microorganisms. For me natural history is not an old-fashioned form of knowing; it comprises the narratives of living entities that provide the alphabet of the design vocabulary.

Each of the four cells in the design has a different but complementary function. Cell #1 is for cultivating fishes; in the above case I used tilapia, a species that grazes on algae "pastures" that are cultivated on screens in other parts of the system. When the screens have been grazed down by the fish, the screens are removed to recover and grow more attached algae in cells #3 and #4. Also, growing in cell #1 are tiny floating plants, particularly duckweed and Azolla. They supplement the diet of the tilapia. Duckweed and Azolla are cultured in other cells as well and harvested frequently to prevent the plants from blocking sunlight getting into the system.

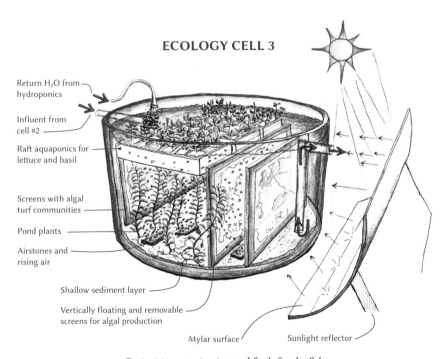

Optimizing growing internal feeds for the fish

ECOLOGY CELL 4

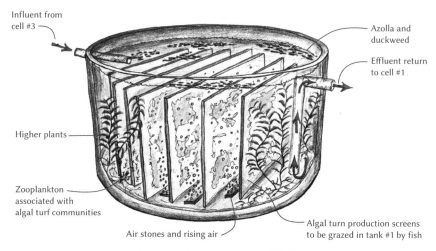

Water purification and internal feeds

An important component in the first cell is the presence of an exotic fish species that helps consume oxygen-robbing solids that accumulate within. In this particular instance it is an armored catfish. Its bottom-swimming behavior stirs up the fish manure and causes it to flow downstream into cell #2. The armored catfish also eat some of the tilapia wastes that settle on the bottom of tank #1; this is the catfishes' primary food. In this fashion, waste products do not accumulate and contaminate the water, allowing the fish to grow rapidly.

Flowing into cell #1 is clean water and a variety of organisms or plankton from upstream in the system. Cell #4 water is highly purified and rich in supplemental living feeds for the fish.

Cell #2 is for converting waste solids to mature bottom sediments as an eelgrass community does. Here I am trying to simulate the processes that take place there with a series of symbiotic relationships that quickly create productive sediments instead of toxic sludge.

This sketch depicts the eelgrass analog. The aquatic plant is the Canadian pondweed, *Elodea canadensis*, a plant with a large surface area. It is a powerful producer of oxygen during daylight hours. Also in the system is an inert nonwoven fabric that acts as an additional substrate for diverse life forms, including attached algae.

By converting the fish waste solids into a mature bottom substrate or sediments, cell #2 with its clams and bacteria develops a biochemical complexity capable of supporting a diversity of life, just as the eelgrass community does.

Cell #3 has a variety of functions that include water quality improvement, fish food culture, and the culture of land plants including lettuce, basil, tomatoes, and cucumbers.

The horticulture is done on floating rafts. The roots of the food crops extract nutrients from the water and provide a substrate for microorganisms and plankton. The tank also contains attached algae screens that are rotated back into the fish culture cell. Note in the drawing the solar reflector to add light as the horticulture "rafts" block some direct sunlight entering the system. At this stage, the Canadian pondweed is still present, but it is no longer dominant, having been replaced by the algae turf screens as the primary food source for the tilapia.

Cell #4 is the ultimate water-purifying component in the system. Its screens host algal turf communities that extract nutrients, support animal plankton populations, and create ultra-pure water. The water from cell #4 flows back to the fish culture tank. (This is shown in the illustration "Water Purification and Internal Feeds.")

Several prototypes were built in a greenhouse located in the Intervale, an agricultural development center in Burlington, Vermont. The following photo depicts several of the systems.

The original species of fish chosen to culture in the systems was blue tilapia *(Oreochromis aureas)*. I had many years of prior experience culturing this species, which was ideally suited to the eelgrass design. The fish were confined to tank #1 and were stocked at an overall system density of approximately one fish per 8 gallons of total hydraulic capacity of water, a density equivalent to one fish per two gallons of water. The water flows from tank to tank on a continuous basis in order to provide purified water at all times to the fish.

Food-growing eco-machine

This density was relatively low by many contemporary standards of fish culture, but we did not know at the time how quickly the system could purify the water or to what extent it could produce the feeds needed by the fish for them to grow rapidly. But grow rapidly they did. Within nine months they were large enough for market. But the real news was in the feed conversion ratio (FCR). This is a measure of how many pounds (dry weight usually) of external feeds are required by an animal to produce a pound of edible food. The norm for aquaculture is an average of two pounds of feed to produce one pound of the desired species. The ratio can go much higher if culturing carnivorous species that feed at the top of the food chain.

Our prototypes produced fish extremely efficiently. The feed conversion ratio was less than one. This seems like an impossible number, the equivalent of a perpetual motion machine for animal farming. For every pound of fish produced, less than a pound of external feed was added to their diet. The difference between a conversion ratio of less than 1 compared to the average of 2 or higher was due to the food for the fish being produced within the system

Tilapia

itself. About 60 percent of their diet was cultivated within. It was behaving like a real eelgrass community with its internal productivity.

The above photo shows the quality of the fish produced.

As I reflect on the importance of seagrass meadows as important teachers for an ecological age, I ponder the nature of other ecosystems like mangrove swamps, for example, where freshwaters, the sea, and the land come together—and the stories they might tell us. They are also very productive ecosystems and nurture diverse forms of life too. What do they "know," that we need to know, to feed the human family in a way that helps heal the planet? And what about coral reefs? There is such bounty in their nutrient-poor environment. Their dynamic nature may provide the clues we need to design anew in an era of information richness and material scarcity. There is so much to discover.

5

Restoring Polluted Waters Ecologically

Several decades ago scientists worried about the global effects of an all-out nuclear war. Their hypothesis was that such a war would create firestorms that in turn would spread large amounts of black carbon or soot into the atmosphere. Their models showed a darkened sky lasting for many years, decades or more, causing a major destabilization of the Earth's ecosystems and with it, untold harm.

We are beginning to witness the equivalent of a "nuclear winter" within the world's waters. It is not being caused by nuclear radiation or firestorms, but carbon is definitely at the heart of the story. Unlike a nuclear winter, this tale has begun and is unfolding now, albeit at a slower pace. Lakes and rivers are browning and becoming murky. They are darker and this browning is blocking sunlight from getting to bottom-dwelling algae and even to free-floating phytoplankton in the waters above. This is reducing photosynthesis, and as a result internal oxygen production is less and in some cases has almost been extinguished, leading to the reduction and elimination of fish habitats. The phenomenon has been chronicled around the world.

Browning is critical in so far as it affects photosynthesis and the production of oxygen. It is not widely known, but close to half of the global photosynthetic carbon fixation is carried out in aquatic environments. In short, these environments are not only important but crucial to a healthy planet. Reversing browning should be at the top of the to-do list of the world community.

Browning of the waters has been caused by a dramatic increase of carbon into the waters. In this case, the carbon is dissolved organic carbon (DOC).

The decade straddling the new century saw a doubling of dissolved organic carbon in the waters of North America and Europe and it continues to increase. The browning is influenced by and is in turn influencing climate change as well. With less light entering the waters, non-photosynthesizing bacteria begin to dominate and feed directly on the organic carbon, producing more than normal amounts of carbon dioxide. The excess carbon dioxide in turn enters the atmosphere and drives climate change even further.

Some scientists have attributed the rise of DOC to a reduction in sulphur in the environment, resulting from declining acid rain levels as industrial emissions are being curbed. With declining sulphur concentrations, the dissolved organic carbon became "unstuck."

While this may be a factor, I believe that the primary cause of the increase in browning may lie in changes in how landscapes are managed, particularly in agriculture. Modern farming and landscape management techniques, with their use of chemical fertilizers, pesticide, fungicides, and herbicides, as well as the practice of constant plowing are breaking apart the ecological relationships that normally keep the various forms of labile and stable carbon in dynamic circulation in the soils. These factors, combined with loss of biodiversity caused by the increase in the scale of deforestation and expansion of monolithic lawn environments, are compounding the problem. Soils systems, essential to the health of the planet, are subject to stress and decline; the carbon content in soils is on the decline almost everywhere. The result is an exodus of dissolved organic carbon into receiving waters during rains. The life that normally keeps carbon in the soils is going, or is gone.

Even a small country, like England, with its relatively benign climate, has lost over 4 million tons of soil each year for the past twenty-five years. On a global scale, soil loss, and with it soil carbon, has been estimated to be at least 75 billion tons per year with an annual economic loss of $400 billion. The figures are almost incomprehensible and their consequences are tragic.

Yet this need not be. Soils can be created and improved. Courtney White in his book *Grass, Soil, Hope: A Journey through Carbon Country* writes of the experiences of successful soil rebuilding experiments around the world. It is a remarkable tale and one with the potential to revolutionize how agriculture and landscape management can be transformed in the twenty-first century. The breakdown of the world's soils can be reversed, even in arid and semiarid areas such as the American Southwest. There are a number of working examples of this reversal already.

Healthy soils are very complex systems with unprecedented amounts of diversity. One of the most significant parts of the whole system is the relationship between the root systems of plants and their symbiotic relationship with beneficial mycorrhizal fungi. The term *mycorrhiza* refers to the complex role of the fungi within the root systems of plants. There are several fungal groups with distinctive root associations with plants, of which the best known are the arbuscular mycorrhiza. This combined plant root and fungi mutualism is one of the keys to soil formation and the sequestration of carbon from the atmosphere.

Including carbon, nutrients from the plant roots in close association with the fine root hairs of plants and the long, branching filamentous structures of the fungi called the hyphae produce a sticky protein. The protein is called glomalin and has been described as one of nature's superglues. Glomalin is a glycoprotein produced on the hyphae, or spores of arbuscular mycorrhizal fungi; collectively they are called mycelium. Eventually glomalin binds itself to particles in the soil and forms larger and larger aggregates that mediate soil formation and carbon sequestration. The soils become resistant to wind and water erosion and a magnet for living organisms. In the process humus, the dark organic matter characteristic of rich soil, is created.

Soils are the greatest repository of carbon on earth. They play a seminal role in climate management. Healthy soils, which increase their carbon content annually through farming ecologically, represent our best chance to bring climate change under control. High-carbon soils are also our best bet to reverse the browning of water bodies.

Through studying how soils and soil carbon work, I have come to see that contemporary methods of remediating polluted waters are flawed by many of the same mistakes that are being made in agriculture and landscape management. The common use of chemicals, such as alum (aluminum potassium sulfate), to control excess algae in lakes and ponds weakens the ability of a lake or a pond to manage its own nutrient- and carbon cycles.

Like most water stewards, I have been guilty of seeing the remediation of waters in a too linear fashion. As a general rule, I have followed Leibig's law of the minimum. Leibig's law states that the potential of a system, such as soil or a water body, is dictated by the amount of the least abundant nutrient: a chain is only as strong as its weakest link. In the case of alum use above, the chemical phosphorus is deprived of its ability to cycle in the system by the alum and it effectively becomes the weak link in the chain.

In my own work with floating eco-machines and other kinds of eco-machines, we consciously use living technologies to remove nitrogen from the water and therefore make nitrogen the limiting factor in the equation.

This is done through a three-stage process. First, the presence of abundant oxygen, which is often pumped into the water body at a significant energy cost, triggers bacterial nitrification. Ammonia is converted to nitrites by one group of bacteria and then by another group of bacteria to nitrates, also through an oxygen-demanding process. Afterwards with the circulation of water through carbon-rich and oxygen-poor environments, the nitrates are converted to nitrogen gas through a bacterial process called de-nitrification. The nitrogen gas in turn leaves the water and enters the atmosphere. In this way, we limit nitrogen and mitigate excessive blooms of algae.

The ecologically based technologies including the eco-machines and their floating restorer counterparts have proven to be effective technologies and many a body of water has benefited from their presence. Despite their successes, I worry about their efficacy in the big picture. They do the job, but we have yet to convincingly prove that they can trigger beneficial processes that will allow for a body of water to self-heal internally, including sequestering its own carbon as can happen in soils. Are we overlooking the carbon cycle and is there the equivalent of glomalin, the protein "glue" in the soil, being made by aquatic plants and fungi in watery ecosystems?

Although the literature is sparse, it has been found that the fungal and plant symbiosis, as well as glomalin, have been recorded from some, but not all, aquatic environments. This is good news if one is attempting to create a carbon-centric rather than a nitrogen- or phosphorus-centric approach to the healing of polluted waters. In other words, what would it take to remove dissolved organic carbon (DOC) from the waters and convert that carbon into stable organic carbon based upon sticky proteins like glomalin? By focusing on the carbon side of the story, we might just change the whole equation and create the aquatic analog of dark healthy soils.

The question arises, what is the best way to create treatment approaches that are holistic, embrace all the kingdoms of life, and equally important, obtain high rates of water circulation between the water body and living organisms, including plants and their root systems? Secondly, our ecological technologies need to be designed to impact well beyond their immediate treatment zones. In other words, they need to function as incubators for beneficial organisms for the whole pond or lake. In an inshore marine environment, such as a bay, cove, harbor, or salt pond, they would have to have a similar catalytic function.

Our hypothetical floating eco-machine would have to be composed of a series of discrete elements. First, the water would have to circulate and possibly be aerated. We have used vertical axis windmills in the past, as well as solar cell–based circulation and aeration. In Flax Pond on Cape Cod we used a hybrid solar electric/wind generator to good effect throughout the 1990s.

Secondly, it will be necessary to continuously feed the technology with beneficial organisms with trace minerals and electrolytes. This mixture should include algae, fungi, and bacteria, trace minerals from rock powders or seaweed, and mineral electrolytes capable of carrying a weak electrical charge.

Thirdly, the technology needs a huge amount of surface area to support the above organisms. Years ago, I created and patented a technology that I called an ecological fluidized bed. Very large volumes of water were airlifted through a semi-buoyant media with attractive microsites for beneficial organisms. Another way to put large volumes of water in contact with surface areas would be to circulate the water through the root systems of plants floating on the surface. The fine root hairs of plants have incredible surface areas for treatment.

Fourthly, the water needs to flow through carbon-rich sediments, the equivalent of soils in lakes and ponds. This is done with the aid of coarse media bottom filters through which the water is circulated. Water treatment in these out-of-sight zones can be remarkable.

Finally, our restorer eco-machines need to be biologically diverse. In some cases, aquatic life from over half a dozen local water bodies contribute to the system. Filter feeders, such as freshwater mussels, are especially important. Plant species like bulrush *(Scirpus spp.)* are also key actors on this ecological stage.

The other key challenge will be for us to create really cost-effective solutions that would encourage communities to willingly adopt an ecological approach to managing their waters. Maybe a step-by-step approach would be worth a try. The first step would be to create water circulation, followed by a second step, namely the addition of beneficial organisms and trace minerals. There is a substantial amount of evidence that the right kinds of supplemental microorganisms of sufficient density can dramatically improve water quality. This includes improved water clarity, the inhibition of algae blooms, nitrification, and de-nitrification, as well as the digestion of bottom sediments.[1]

Depending upon feedback, if the system is not responding dramatically, it might prove wise to ratchet up the process to a third step that would include high-rate restorers with both aerobic and anaerobic treatment components. Of course, each step would increase the treatment cost, hence my step-by-step approach. The following illustration shows how it is possible to combine floating parks with the restoration of polluted waters. They would work in freshwater and inshore marine environments, including urban harbors.

Coming back to my larger concern, namely the healing of waters and making ponds a sink for stable carbon, I believe it is time to undertake trials in a variety of aquatic settings. While we do not have a lot of evidence to go on yet, we do know that healthy ponds and lakes can be carbon sinks, whereas degraded ones are excess exporters of CO_2 into the atmosphere. One area that interests me is the development of biological incubators that produce beneficial organisms in huge numbers, which can be released in an ongoing basis into lakes to support the healing processes. If we can combine water remediation with aquatic carbon sequestration, there is a good chance that we will find ourselves on the front line of helping stabilize the earth's climate. I am committing myself to exploring this frontier.

FLOATING PARK RESTORERS
Can provide an array of overlapping environmental and societal benefits. There are many possible configurations with varying costs and benefits. Floating parks can be linear paths, networks of rings, or combinations fitted with park hubs. The elements are modular, moveable, and expandable as resources allow.

LINEAR PATH WITH RINGS
A floating walking path linking restorer rings and enhancing pedestrian transportation and recreation opportunities around waterfronts or bay areas. Dockage can be incorporated and swing bridges can be inserted to accommodate boat traffic.

HEALING PARK HUB
A water improvement platform and true floating tank. This concept includes twelve picnic areas, mixed dockage, and sanitary facilities. It is reached by boat or the floating footpath. The building doubles as a solar power station with photovoltaic roof and battery bank. The station powers lights, emergency call boxes, and air compressors. The air compressors drive air lifts that move the water across plant roots while aerating the water. This digests sludge and removes nutrients through plant harvesting.

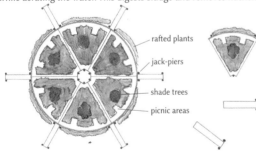

rafted plants
jack-piers
shade trees
picnic areas

FLOATING PARK HUB MODULE
Buoyancy and structure are achieved with a frame of sealed large-diameter HDPE pipe. Walking areas are decked over and plants are held in net.

FLOATING DOCK SECTION
can have plant rafts fixed to either side, creating a lower-cost linear park.

HUB LAYOUT
The hub is composed of six wedge-shaped modules. They are arranged in a circle here, which provides a stable point to support the building, but options are boundless; to have placement, reference local conditions.

ADAPTABILITY
Individual units have small jack-piers, similar to jack-rigs used in ocean drilling platforms, but much smaller. Each pier can be independently raised or lowered to the bottom, jacking the structure up. This allows the sections to be secured and adjusted to local terrain when deployed. They can be small because they are for anchorage more than for the structural stability of the platform. When piers are lifted, the unit is released to float freely. Small tugs can move sections for servicing or to set in a new location.

swing bridge

Restorers as floating parks

6

Healing Degraded Streams and Rivers

I walked several miles to and from school every day as a boy. For me the journey was better than the schooling. My route followed one or another of two streams that ran parallel to each other. They both flowed into Hamilton Bay at the very western end of Lake Ontario in Canada. One of the streams discharged a quarter mile east of our family's house on the shore, while the other entered the bay at an equal distance to the west.

At the mouth of the eastern branch was a cattail marsh with channels that meandered through a maze of tall plants. Heading on inland the stream passed through a culvert that ran under a shore road then climbed slowly into rolling hills. The stream defined the western boundary of a golf course. Trees were sparse. The stream dwindled to become a brook then diminished again as it entered a canyon that led to a bluff from which the stream originated. Up on the bluff the land was flat, and there were intensely managed market farms with a number of greenhouses. The fields and service lanes shaped my course as I crossed the fields to the main road that linked the city of Hamilton with the town of Burlington. My school was on the far side of the road on the edge of a suburban area.

The stream to the west was completely different. Its discharge was measured and steady. It passed through a marsh and bog complex then meandered in dark shade up a narrow canyon with steep slopes until it reached the base of a bluff. The slopes were heavily wooded. The stream's origins were a series of springs. Skunk cabbage, Jack-in-the-pulpits, and marsh marigolds were well established along its lower reaches. Watercress was abundant. On top of this bluff was an agricultural area, made up mostly of orchards that grew cherries and peaches. Further inland, beyond my school, the orchards on the limestone escarpment mostly produced apples.

My choice of routes to a large degree was determined by whim and time. The eastern or golf course route was a few minutes faster. As I got to know the streams, I began to see them as having very different characteristics. In anthropomorphic terms the eastern stream was violent and erratic. It was dirty and filled with silt and prone to both flooding and drying up. Apart from spawning carp in the lowest reaches, there were no fish and bottom life seemed nonexistent. There were no snails, frogs, or even caddis fly larvae. I spent hours looking and waiting for minnows to appear. They never did.

The western stream, despite being similarly sized and in the same watershed, had a totally opposite nature. It flowed steadily through wet years and dry. Its waters were crystal clear and its temperature remained steady. It shone in dappled light and darkened in deep shadows. It gave off organic odors I still can smell to this day well over half a century later. The big hardwood trees and understory plants imparted to me a sense of jungle-like bounty. Best of all, the stream hosted many creatures that I came to know. Most remarkable to my childhood self was the spawning of the fish in the spring. Just after the flowering of the marsh marigolds, suckers—fish with big, somewhat ugly-looking sucker mouths—entered the stream in large numbers. They moved upstream into the darkest reaches and thrashed about in sexual orgies, oblivious to my wading amongst them.

More than anything it was vitality of the place that attracted me; an essential aliveness that I began to understand in later years. The place spoke to me. The western stream and its valley, despite their modest size, were an intact ecology. It contained the remnants of the wild. The other stream was neither wild nor a whole ecosystem. The deforested landscape could no longer absorb and hold rainwater. The soils had been washed away and the springs had died. There was an elemental sadness about the stream. Its moods were a byproduct of its exploitation, followed by neglect then decline. The golf course there for me became a symbol of the theft of wildness.

A few years later, my father, sensing my deep unhappiness at being surrounded by so much development and biological destruction, found for me a series of books that changed my life. They were Louis Bromfield's *Malabar*

Farm books. They told the story of his return from Europe at the outbreak of World War II to an Ohio farm and his settling there in a worn-out agricultural landscape. Using ecological methods, combined with farming practices learned from French peasant farmers, in just over fifteen years he transformed the hilly landscape into a bountiful Eden.

I was mesmerized by his ability to build the equivalent of hundreds of years of topsoil within decades. His integration of animals with plants seemed almost revolutionary. What struck me as most miraculous were his tales of the return of the springs, springs that years before had dried up following abuse of the woods and overgrazing the fields. What Bromfield taught me became my greatest lesson and my most enduring gift. At thirteen I learned from him that ruined lands could be restored and discovered that in the teachings of ecology lie the foundations for healing, not just individual landscapes, but possibly the whole world.

Ecological wisdom is learning that life connected to life across great spans of evolutionary time. It is about complex relationships between species and their environments, and about self-organization, self-design, and self-repair. It includes the perpetuation of diverse systems that are constantly co-evolving. It is an elusive dance that one can sense but never fully understand. It is real nevertheless.

Ecological scientists are chronicling the declining fabric of the Earth. The term they use for this is *trophic downgrading*,[1] meaning the elimination of essential species that provide management for large and complex ecosystems. Our planet is currently undergoing a mass extinction caused to a large degree by a single species, *Homo sapiens.* With our insensitivity to the workings of nature, we are pulling apart the Earth's fabric through deforestation, extractive farming, industrialization, and wholesale slaughter of the iconic animals of both land and sea.

Ecologists have discovered that the structures and organization of ecosystems are determined by a top-down control over the life forms contained within them. This is usually done through the feeding of the predatory animals at the top of the food chain. For example, sea otters enhance and protect kelp abundance by feeding on kelp, eating herbivorous sea urchins. When sea otters are killed, the kelp will be overgrazed by rising populations of sea urchins and will subsequently disappear. Healthy kelp forests are a major sink for atmospheric carbon and thus help maintain essential climate balance for the whole planet.

When large-mouth bass are present in cool freshwater lakes, they increase water clarity by consuming smaller fish that feed upon tiny zooplankton, which normally reduce algae in the water. When bass are removed from the system, the small fish increase, the zooplankton decrease, and algae populations can explode, causing serious decline in the lake.

On coral reefs, it has been found that the overfishing of sharks and large reef fishes causes a decline in healthy reef-building corals and coralline algae. The presence of the alpha predators keeps the coral-based community intact. There are many, many examples of trophic downgrading, as a result of humans eliminating essential predators at the top of food chains. The same phenomenon has been observed in terrestrial ecosystems. Wolves, for example, can control and reduce elk populations. Smaller numbers of elk result in less browsing of shrubby plants that can result in an explosion of scrubby vegetation. Grasses subsequently decline and along with them many species of herbivores that graze on them.

Trophic downgrading has widespread ramifications beyond the loss of species and ecosystem structures. Wildfires become more common, for example. Additionally, there have been discovered a number of links between increases in diseases and the loss of top predators. Physical and chemical processes within ecosystems also can be altered. For example, trophic cascading in lakes can shift them from being net sinks to net sources of atmospheric carbon dioxide.

There are innumerable examples of the breakdown of ecologies around the world. Many have been ruined before they have been studied. The wholesale decline of streams and rivers represents a major case in point. Over the past century, fossil fuel combustion and the use of agricultural fertilizers, as well as poor treatment of wastewaters, have caused the rate of nitrogen input onto the landscape to more than double, killing rivers worldwide.

Oxygen, essential to the life of streams, disappears, because of excess nutrients disrupting normal food chains. Bacteria and boom-and-bust algae cycles cause the normal biodiversity in the streams and rivers to disappear.

Degraded Plankenburg River in South Africa

Biodiversity and *niche diversity* keep normal streams healthy; niche diversity is the role an organism plays within a community as well as the specific microhabitat it inhabits.[2] It has been shown that biodiversity and niche partitioning can improve water quality, done in part by a more efficient removal of nutrients and pollutants. This is an extremely important piece of information for restoration biologists. Creating and maintaining biodiversity and distinct niches can become a tool for ecosystem improvement. How can this be done is the central question? Are there ways to engineer systems ecologically that might restore health to a stream degraded through loss of species and habitat?

It has been argued that the only way to restore such a stream is to get rid of the toxins, fertilizers, and excess silt by going upstream and stopping the offending practices at the source. This is certainly true as far as it goes, but I long ago observed that in most instances around the world such a task is close to impossible. In South Africa where we work, the slums are often situated on the edge of streams. In such communities, sewage often runs down the streets and enters the streams along with the garbage. Streams in this kind of situation do not have much of a chance to remain healthy. The life within them dies.

There is an alternative approach, but it involves a course of action rarely taken. I am referring to the practice of directly intervening in such a stream and introducing ecological technologies to trigger and facilitate the internal healing of the ecosystem. Might we reverse trophic downgrading in streams, replace it with trophic upgrading, and take steps to create habitats that critical species will find hospitable? Going one step further, we should ask whether beneficial niche partitioning can be reintroduced into a body of water and learn how to do so.

I don't yet have the answers, but I do have some potential solutions based upon my experience with designing and operating eco-machines. Hopefully these suggestions will lead to a proliferation of ecological intervention techniques for streams. Several potential solutions are described below.

RESTORING THE RIVERS: ideas for bringing riverine ecosystems back from the brink that are technically simple yet biologically complex. These examples offer a range of complexity, appropriate for different context, and all scaleable.

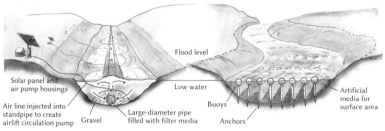

1. ECOLOGICAL PIPELINE
Airlifts aerate water and circulate it through sediment pre-filters and then through media in the pipeline and the associated surface ecologies.

2. ARTIFICIAL KELP FOREST
Artificial media are anchored to the bottom and float to the surface in long treatment ribbons. Water contacts the organisms colonizing the media. Zero electricity required.

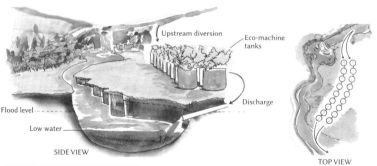

3. ECO-MACHINE RESTORER
Passive inflow from upstream diversion. Two parallel lines of light-transmitting tanks allow for internal photosynthesis and the potential for zero electricity requirements.

Three possible living technologies for stream and river restoration

ECO-TECHNOLOGY 1: AN ECOLOGICAL PIPELINE

The function of the pipeline will be to filter stream water and reduce levels of pollutants in the water. The processes will be microbial with bacteria and possibly fungi playing dominant roles. If water clarity is improved, as we hope, it will allow for more light to penetrate the water, thereby enhancing the photosynthesis by attached algae or algal-turf communities. Should this take place, the stream will experience increased biodiversity and support aquatic insects, snails, and fishes.

This concept involves a pipeline that runs along the bottom of a stream. The pipe would be perforated, filled with a coarse media, and covered with gravel. Even at low water levels during the dry season, the pipeline would be covered. The stream water would be aerated and circulated through the interior of the pipe and the surrounding gravel. The system would be invisible insofar as no physical structure would be obvious from the stream bank. The length of the prototype pipeline should be between one-quarter and one-half mile in length.

There is one drawback compared to the other schemes that follow: the pipe will require electricity to power the air compressor. I propose that solar power should produce the electricity from a solar power station situated above the stream's floodplain.

A major technological challenge will be water circulation through the media along the length of the pipe that will use airlift technologies. We have used a number of approaches over the years, but never quite in the way proposed here. A short segment model will have to be built and tested, initially in a tank or a shallow pool. One of the benefits of the ecological pipeline approach is that the technology will be relatively immune to high water and flooding conditions.

ECO-TECHNOLOGY 2: AN ARTIFICIAL AQUA-FOREST

For years, we have used artificial fabrics to provide surfaces for beneficial organisms that aid in the purification of waters. Remarkably high numbers of life forms can occupy somewhat restricted spaces with such media present. The primary purpose of the Aqua-Forest concept is to insert biologically friendly surface areas into the stream environment on a large scale. This concept involves ribbon-like structures that are anchored on the bottom at the upstream end of the treatment area. They are no deeper than the shallowest zone of the stream during the dry season. (See illustration "Three Possible Living Technologies for Stream and River Restoration.")

This is the simplest of the schemes. It requires no electricity and, apart from the anchors, no infrastructure beyond the media. In the past, we have used nonwoven fabrics of the kind often seen in car wash facilities. Some of the fabrics are quite sticky and optimal for the attachment of microbial and algae-dominated communities. Another benefit of the Aqua-Forest approach is that it is easily scaled up and flexible enough to allow for a series of distinct sections along the course of a river with each section adjusted to the varying shapes and depths of the stream.

With the attached communities in place, the first benefit to the stream will be an increase in water clarity and a decrease in pollution load, leading to an increase in life-giving oxygen levels. Although to my knowledge the concept has never been tried in a stream, it has been widely used inside eco-machines, and an Aqua-Forest has the potential to develop considerable water purification power and to be the foundation for expanding niche and species diversity appropriate to each stage.

ECO-TECHNOLOGY 3: ECO-MACHINE SOLAR RESTORER

Parallel rows of clear-sided tanks would be placed above the floodplain parallel to the stream, as shown in the illustration. Natural light is transmitted through the walls of the tanks into the interior. Through the photosynthesis of the microalgae, during daylight hours these systems can produce significant amounts of oxygen. The tanks are connected to each other with piping that forms two parallel streams separated from the main body of the actual stream. However, stream water will flow into the system at the upstream end and the outlet water will rejoin the stream. The life within the tanks will clean the water flowing through the system. By the end of the process the stream water will be clear and oxygen-rich.

One of the advantages of the system sketched above is that it can operate without electricity as flow through the system is by gravity. At some locations, a disadvantage of the eco-machine restorer might be that it would need some kind of fencing or protection from vandalism.

The eco-machine restorer could require ecological elements to purify the water that would tend to be different at various stages along the treatment process. Polluted water at the input end would be exposed to a unique assembly of organisms for treatment. The first third of the tanks would house artificial kelp substrates in the form of curtain-like structures made from water-absorbing media that have been inoculated with bacteria and algae to constitute part of the functioning of the early stages of the system.

The middle group of tanks would have floating racks on which marsh plants and water-loving plant species would grow. Their roots would penetrate down into the water and provide surface areas necessary for diversifying the biological communities. They would include water-filtering animal plankton and freshwater clams and mussels.

The final series of tanks would be seeded with species from streams normally found in clean waters. This community would have fishes of a diversity of small species as well as top predators. These tanks would also have rafts with higher plants growing on the surface.

Eco-machine for purifying stream water in South Africa

One of the great advantages of the eco-machine restorer is that it would become a destination and a focal point for educational and stream restoration activities. Another advantage would be the visibility of many of the organisms, enabling people to come to appreciate the contributions of the different kingdoms of life in supporting stream health and healing. These could include riparian tree and shrub seedlings for subsequent stream bank restoration projects. One of the functions of the surface plantings would be to protect the fishes from bird predators such as kingfishers. I once had the misfortune of losing fifty young bowfin fish to kingfisher predation within twenty-four hours because the fish lacked protection from above.

Water leaving the restorer and entering the stream would be clean and rich and have a diverse biota. My hope here is that there would be a zone of clean water and that this oxygen-rich zone would slowly expand its influence downstream.

This eco-machine approach to stream remediation has been tried recently in the Western Cape Province of South Africa near Stellenbosch. The stream was heavily contaminated by waste from slums and agricultural runoff and was unfit for most uses due to excess bacterial contamination. A small eco-machine was built to treat the river water before releasing it back into the river. As of this writing I have very little information on its performance to date, but I am pleased to report there has been a dramatic drop in bacterial contamination in the outfall of the eco-machine.

In conclusion, these ideas need to be tried out in a variety of streams with a range of pollution problems. They can all be tested throughout the world. In some countries like Mexico and India where most streams are conduits for waste disposal, the water is no longer viable biologically. Ecological engineering solutions could reverse this situation, making streams everywhere healthy, bountiful, and beautiful again.

7

The Early Evolution of Restorer Eco-Technologies

A distressing number of ponds, lakes, bays, canals, and lagoons in North America are heavily contaminated with pollutants of all sorts. These include heavy metals, agricultural fertilizers, and poisons, as well as pharmaceutical chemicals including endocrine disruptors harmful to aquatic life. As a consequence, some bodies of water have become biologically comatose, meaning the water is devoid of oxygen on the bottom and sometimes higher. This can result in the loss of communities of the macro-benthos, bottom-dwelling creatures, visible to the naked eye, that are essential to the health of aquatic ecosystems.

In the late 1980s, while working to clean up landfill wastes at the disposal area in Harwich on Cape Cod, I became aware of the state of nearby Flax Pond about a hundred yards to the south. It had been closed to fishing and recreation due to the influx of toxic materials and human pathogens. These were flowing underground into the water at a calculated rate of seventy-eight thousand gallons a day (295 m^3/day). Hydrological studies had concluded that this contamination would continue unabated for decades into the future.

Besides low oxygen content there were high coliform bacteria counts as well as excessive sediment buildup and organic pollutants including volatile organic compounds (VOCs) in the fifteen-acre pond. Compared with other local ponds, Flax Pond's sediment phosphorus (TP) was three hundred times higher than normal. Its iron levels were eighty times higher. This meant that healing the pond, if it were possible, would have to involve developing processes to give it self-healing properties. At that time, there were no off-the-shelf solutions.

Where were we to begin, considering the physical composition of the pond? Measurements revealed an oxygen-rich layer of water near the surface and an oxygen-poor zone close to the bottom. If we could create an internal upwelling to lift water off the bottom toward the surface, we thought it might be possible to de-stratify the pond and expose the bottom water to atmospheric oxygen.

The solution had to be based upon renewable energy as there was no adjacent electricity. Years earlier at the New Alchemy Institute we had solved a comparable oxygen problem in a greenhouse-based aquaculture system by using a small vertical axis windmill mounted on the roof. The blades rotated a shaft made of metal piping that extended down into the water. At the surface of the water was a device made from a large steel baker's bread-mixing bowl through which the shaft passed. When the windmill turned, the mixing bowl spun round and water from the bottom of the tank was pulled up from the bottom and entered the bowl through holes in the shaft, then sprayed out through a ring of holes near the lip of the bowl. When the wind was blowing, this allowed us to aerate the pond. Because the shaft and the bowl presented very little resistance to the turning motion, the device worked well even at low wind speeds.

At Flax Pond, we sought a more ready-made solution. I remembered that in the Midwest cattle ranchers often use floating vertical axis windmills to create water upwelling in their stock ponds to keep them from freezing as the warmer, denser bottom water rises to the top and melts the ice. A rotating windmill shaft turning a blade produces the upwelling; the blade is suspended several feet down in the water column. Its shape was reminiscent of a lawnmower blade and about the same size. They worked very well for us, and I am sorry to report that these windmills are no longer manufactured.

We installed three of these windmills in the more polluted eastern end of the pond. They proved to be reliable and effective at mixing the water adjacent to the devices. To our great delight, we found that oxygen levels in the vicinity began to climb. The effect however was not sufficient to influence the overall ecology of the pond, at least not in the short run.

Something more was needed, and that was living organisms. When it comes to microorganisms, biologists tend to cluster into two camps. The first usually includes ecological biologists, most of whom think and deal in fairly long time frames. They generally see microorganisms as ubiquitous. Given enough time, nature will fill vacuums with organisms adapted to changed environments. In our case, nature transformed an environment previously devoid of oxygen to one rich in oxygen through the action of the windmills aerating the water.

There is, on the other hand, another school of biologists, often those concerned with applied problems such as sewage treatment or solids conversion like transforming garden and household wastes into finished compost. To them the concern is with short time frames, specific species of organisms, and their numbers. Rapid population growth and a high density of organisms are two key objectives for this specificity school.

I tend to sympathize with both the ubiquitous as well as specificity schools. Normally I like to seed an ecologically engineered system with a huge diversity of microorganisms collected from a wide range of local environments. But sometimes conditions do not permit the rapid establishment of a population of the most beneficial organisms and something more is needed. At this juncture in the Flax Pond project, I turned to a prominent bacterial specificity expert for help.

Dr. Karl Ehrlich is a biologist I have known for a long time. In the idyllic countryside of the Eastern Townships in southern Quebec, he has manufactured beneficial bacteria for over thirty years. His products are shipped all over the world to improve the performance of waste treatment and water quality in lakes and ponds as well as to support fish and shrimp farms.

Dr. Ehrlich suggested that we develop a technology for Flax Pond to inject a community of bacteria into the sediments below the windmills. He provided a diversity of organisms that included species of bacteria from half a dozen genera. Most of them had originated in soils. Given enough of them these microbes can digest organic matter and convert ammonia toxic to most aquatic animal life into less harmful nitrates. These nitrates, in turn, can be utilized directly by algae and plants.

The windmill and the bacteria in combination worked well in the immediate vicinity of the windmills. Oxygen levels improved, nitrification or conversion of ammonia to nitrates was detected, and we had some evidence of sediment digestion. However, the impact remained local and we realized that the over 28 million gallons (106,060 m³) of underground liquid waste from the landfill entering the pond annually were overpowering our windmill-bacterial technology. We concluded from the evidence that combining windmill water circulation with bacteria injections into the sediments was a good way to treat mildly polluted ponds. Ponds like Flax Pond, however, which were exposed to large volumes of high-strength wastes, required a more powerful technology to bring them back to life again.

Our design team met to discuss the next steps for the pond. We drew up a list of attributes for our restorer technology. The list included:

1: The restorer must float;

2: It needed to be able to be movable and allow for re-anchoring;

3: The restorer had to be powered by solar and/or wind energy;

4: It had to circulate large volumes of water from the bottom up to the restorer;

5: The restorer must have cellular design:

6: It had to support anaerobic as well as aerobic communities;

7: All the kingdoms of life had to be represented;

8: The restorer needed to support a diversity of plants including shrubs and small trees;

9: Mollusks, especially freshwater clams, had to be its primary filtration units;

10: Diverse rock minerals needed to be introduced into the early stages of the treatment process.

We incorporated all the attributes above into the technology.

The original restorer was built of wood and operated successfully for a decade, beginning in 1991. It was composed of nine cells. The first three cells were designed as ecological fluidized beds, a technology I had patented. The total flow into the system passed through these cells. Restorer 1 had a maximum throughput of one hundred thousand gallons a day. The wind and solar power units did not operate continuously, however, and the system

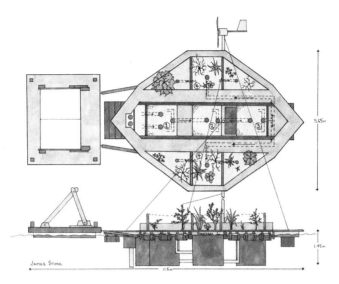

The Flax Pond restorer

averaged thirty-five thousand gallons per day. This was equivalent to 1/500th of the volume of the pond passing through the restorer on any given day.

There were six other cells downstream in the system. The flow was divided into two parallel three-cell components. These housed densely planted water-loving marsh plants, shrubs, and trees mounted within support racks on the surface of the water. The plants grew a huge mass of roots that hung down into the water. The roots supported large populations of microorganisms that helped filter and purify the water passing over them. The colonization of life onto the restorer turned out to be a fascinating process. The communities included freshwater sponges and other colonizing species that added their own filtration processes to the restorer's operation. There was one strange creature that initially looked like an inert blob. It positioned itself where the water current created by the restorer was strongest. It grew and grew and was alive and pulsating with life. It eventually grew to more than a foot in diameter and was made up of countless cell-like structures. We identified it as a colonizing animal known as a bryozoan. Most bryozoa species live in the ocean, but this was a freshwater species. The restorer played another significant role in the ecology of the pond as it became a magnet for young fishes seeking protection and food among the roots.

The restorer

Bryozoan

We installed the restorer in the pond in the autumn. We began to see positive effects by the next spring. We detected and tracked the process of nitrification by measuring ammonia and nitrate levels entering and leaving the restorer. Nitrification is the process whereby *Nitrosomas* and *Nitrobacter* bacterial species convert ammonia to nitrites and then into nitrates, which are a less toxic form of nitrogen. Water entered the restorer at an average of 2 milligrams a liter of ammonia and left at a quarter of a milligram a liter. This was a significant change that was equivalent to half a pound per day (245 grams/day) of nitrogen being converted by the system.

However, the biggest surprise was detected in the pond beyond the restorer. To the best of our knowledge, the comatose pond had not been nitrifying prior to our project, either because the appropriate bacteria were not present, or there was something in the water inhibiting the process. Nitrification is known to be vulnerable to inhibition from inadequate alkalinity as well as from excess organic matter and a diversity of chemicals and metals. However, the process of nitrification began to spread outward from the restorer and eventually throughout the pond, where it had a significant impact on the whole body of water. The restorer had improved the water quality enough to allow a dynamic process within the pond to start and spread beyond the physical limits of the technology. We speculated that it was acting as a bacterial incubator and producing huge numbers of beneficial organisms that spread throughout the pond. From then on Flax Pond began to improve. Within two years, sediments throughout the pond had diminished by twenty-five inches (63.5 cm) in depth, a significant achievement by any measure.

Within six years of the project's inception, Flax Pond was declared open to fishing and recreation. Human pathogens were no longer found in water samples and the fish were declared edible. The pond looked healthy and biodiversity had increased significantly over the years. Also, the town had stopped dumping septic tank wastes into open pits nearby and had capped the landfill so that rainwater did not flush the toxins toward the pond.

Ten years after the project's inception my son Jonathan and I returned to the pond to dismantle the restorer whose useful life had ended. As we hauled the last pieces ashore, we reminisced about the technological problems we had to overcome with the energy and electrical components of the restorer. I felt in awe of the ecological forces we had observed at Flax Pond over the past decade and in other places all over the world (the restorer technology has now been widely adopted). We both sensed that we were seeing into the eyes of the wild and beginning our life's work there.

8

Lessons from the Sea

I have a great affinity with water. I don't know whether it is inborn, the consequence of a childhood living in a house a stone's throw from the sea, or inspired by my father's passion for boats. I do know that the sounds of waves lapping on the shore and the grinding of ice during the spring breakup were true music to me. My imagination also was fueled by the cargo ships that passed near enough to my bedroom for me to see the crew on deck. On occasion, especially during storms, ocean freighters and large Great Lakes ore carriers would anchor very close offshore. As a child I couldn't sleep for the excitement of it.

Water was not only an intimate medium for a kid to explore; it was my imaginary highway to the world. Sailing alone in my little Sabot pram dinghy, I daydreamed happily, seeing myself casting off from our backyard dock, heading east the length of Lake Ontario, down the St. Lawrence River to the sea and finally out into the wider world. When I was about ten or twelve, I would hang out in a local boatbuilding shop just down the road. A pair of double-ended Tahiti ketches of the type designed by Colin Archer were being built there. I thought they were the most beautiful things in the world. They were designed to sail to the South Pacific and beyond, and I wanted to go along. In our extended garage my father was building a sloop designed by naval architect William Atkin. I reckoned that if I played my cards right, in a couple of years I might be able to sail it out onto the great watery highways to faraway places.

Water remains my medium. These days I sail on Nantucket and Vineyard sounds in a beautiful twenty-nine-foot Herreshoff ketch that used to belong to my father. Mostly I work to restore polluted waters and to clean up various kinds of liquid wastes. I am lucky in that I have been able to visit and work in some of the places I dreamed about as a child. These include the Caribbean, the island of Tahiti in the South Pacific, and the Seychelles in the Indian Ocean. In these places, I developed aquaculture systems to purify and recycle water. I also had the memorable experience of sailing a thirty-three-foot long three-hulled vessel or trimaran, offshore to South America.

The meaning of water continues to possess my consciousness. The fact that we humans are close to 70 percent water has convinced me that water is important to even the most landlocked among us. For years, I have suspected that the quality of water we ingest affects many things about us, as well as the plants and animals upon which we depend. I work under the assumption that water quality helps determine our quality of life. If we despoil it, it will be at our own peril. Nothing I have learned in recent years causes me to doubt this.

My associates and I at Ocean Arks International, our not-for-profit, and our design firm JTED design build living technologies that are used to treat wastes and to restore polluted waters. Our experience includes improving fresh and brackish ponds and cleaning up contaminated canals as well as installing sewage treatment in both China and here at home. We call our technologies eco-machines and our floating technologies restorer eco-machines. They are ecologically engineered, by which I mean that they are designed with the attributes of natural ecosystems like marshes, ponds, and streams. In a sense, such natural ecosystems are like parents from which we borrow their designs and life forms. Eco-machines contain representative species from all the kingdoms of life. These range from bacteria and viruses at the lower end of the size scale to fungi, animals, and woody plants at the macro scale. Working together as a biological team, these assemblages of organisms help us transform polluted water into clean water. In return, we provide them with extra air and water circulation as well as appropriate substrates to live on or in.

In our case the engineering is human, but the cast of creatures that actually do the work is gathered from a diversity of wild systems. Within the eco-machines such teams organize themselves into new and complex living systems. In this sense, an eco-machine is a partnership between human designers and other life forms with the objective of solving problems.

As a designer, I face one basic question: namely, what models should I use to obtain the instructions I need to create such environmental technologies? In short, what is my inspiration for design? The model always comes from one natural system or another. For example, if I am using a pond as my primary model for a technology, then the architecture of the pond combined with its various functions, as well as the life within it and the dynamic nature of the system itself, all provide me with clues and ideas. The same would be true if I used a salt marsh as my model. However, a pond-inspired living technology would be different from a salt marsh technology.

The more I look to the nature of ecosystems for inspiration, the more I have come to appreciate the countless subsidies from the natural world that sustain us, producing the air, water, soil, and bounty that makes human society possible. The scales have fallen from my eyes as I have studied the ecologies that nurture us. For example, when I first discovered marine eelgrass, I considered it as simply the stuff that caused my anchor to drag or fouled my lure. Then I came to see it as a resource that washed up onto the beach which, when I put it on my garden to fertilize the soil, would give my family tomatoes. Later I started to use eelgrass as bedding for our chickens. The hens loved to feast on the myriad of small shrimp-like creatures associated with it. These beach fleas in their diet turned our hens' egg yolks sunset red, and they were much in demand by my family and friends.

Eventually and very slowly I came to realize that the eelgrass community is one of the keys to the bounty of our inshore waters. What happens in these communities, in some sense, mirrors what will happen to us. Their health and continuity in the waters of coastal New England matter to each of us, even if we don't venture out onto the ocean.

Eelgrass, scientifically known as *Zostera marina*, is an aquatic flowering plant, not a true grass. It relatives are members of the freshwater pondweed family. Unlike the seaweeds, eelgrass puts down roots into sand and mud. They are tall thin plants that grow very densely; up to three thousand plants can grow on a square yard of bottom-forming meadows below the low tide level down to a depth of fifty feet.

These days I am trying to learn more about such shallow water communities. They have become the inspiration for the design of new water-purifying and marine aquaculture technologies. Eelgrass communities are superb at protecting and restoring coastal waters. Yet they are also vulnerable to human abuse. Their gift to us is that they can be superb teachers.

Eelgrasses are water-quality workhorses of the first degree. As they wave back and forth in the currents and tides, they slow down the rate at which the water travels through them by half. As the water slows, fine silt and other suspended materials begin to settle onto the plants and on the bottom. Consequently, water clarity goes up and the settled material creates rich bottom substrates that support a productive and diverse food chain.

A second attribute of eelgrass is unusual in the plant kingdom. For some reason, very few creatures eat it. It doesn't seem to be palatable. This protection from direct grazing is a critical feature of the community and a key to understanding its importance. True, Brent geese nibble it near the low tide mark and sea urchins may graze on it, but in the main this plant is left alone. Eelgrass can produce huge and relatively stable surfaces that can be colonized by organisms looking for attachment and protection. For every square yard occupied by eelgrass on the bottom, the plant has twenty times this surface area in the water column above. This represents some twenty-six thousand square inches of plant surface available to be colonized over each yard of bottom. This is natural engineering at its best. Eelgrasses are not only water purifiers, they are food chain builders without compare in the natural world.

Many other plants attach themselves to eelgrass, including a vast number of algae and seaweeds. Besides the algae, there is a wide diversity of water-filtering creatures who depend on eelgrass environments, such as sponges, stalked jellyfish, anemones, hydroids, tunicates, and sea squirts. The range of life on eelgrass in unpolluted waters stretches one's credulity. These life forms are hard at work filtering the water, removing nutrients as well as low-level pollutants, and providing lunch for the legions of microscopic and macroscopic grazers and predators.

Down on the bottom, in the rich sediment deposited by the eelgrass, is a motley crew of scavengers. They constitute a recycling department that is very well run. The largest member is the horseshoe crab, but the blue crab tends to dominate. Hermit crabs are often numerous and spider crabs, as well as mud crabs, are part of the team. When observed under a hand lens or microscope, the smaller sediment-burrowing creatures look like creatures out of science fiction, but their roles in the overall ecology of the community are essential. One member of the bottom-dwelling community, beloved by humans, is the bay scallop. The fate of the bay scallop is tied to the fate of the eelgrass. The blue crab may dominate, but the scallop is the totem animal of the ocean floor.

There are other benefits engendered by the eelgrasses. They literally feed other organisms. During photosynthesis, just under 10 percent of the organic carbon fixed by the eelgrass becomes food for other water-purifying organisms, including beneficial bacteria. The eelgrass and associated algae also provide beneficial gases. During the day, when many fish come to them for cover, these plants provide additional oxygen through photosynthesis. The water in the seagrass bed becomes saturated with the life-giving gas.

Eelgrass

Scallops on eelgrass

The water-purifying ability of the eelgrass is matched by its productivity. The eelgrass and attached algae are about equal in overall mass or standing crop. The heavily grazed attached algae may be the more productive partner of the two, but this bounty is based upon support from the host plant. The exact productivity of eelgrass communities is still the subject of scientific controversy. There is no doubt, however, that these seagrass meadows are among the more productive areas in the sea and rival industrial agriculture in output. If one adds to such productivity the sheer diversity of life found there, we are talking about a very special environment, one that shapes much of the life and the waters along the coasts of the world.

Eelgrass communities are not well known for their contributions to water quality, but are famous for the fish they host. The list of fish that visit eelgrass areas for food and shelter during some part of their life cycle reads like a who's who of commercial and sport fisheries. They are magnets for species we like to eat. Species with no commercial value also thrive in eelgrass; my favorites

Seahorses

are the pipefishes, seahorses, and sticklebacks. With fins, goggles, and snorkel I can stare endlessly at these clowns of the sea, male seahorses with babies in their belly pouches, pipefish drifting like flotsam amongst the grasses, and the sticklebacks with their darting movements and flashes of color.

Sticklebacks

HOME-SCALE SHRIMP AQUACULTURE ECO-MACHINE

This system works nicely to convert food waste into delicious seafood. This assemblage is composed of species tolerant of a wide range of salinity, making management of the system much easier. Using inexpensive, light-transmitting algae culture tanks, this recirculating system is beautiful and fun. Those in northern climates can place the tanks in a small greenhouse.

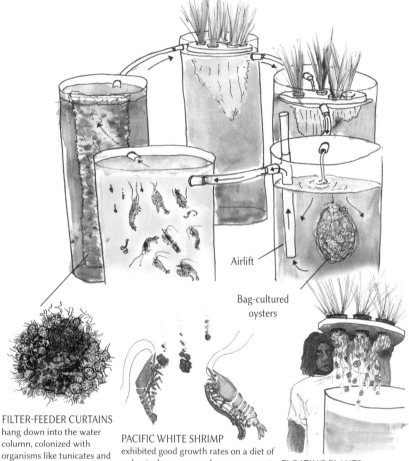

Airlift

Bag-cultured oysters

FILTER-FEEDER CURTAINS hang down into the water column, colonized with organisms like tunicates and sponges that actively filter water. A huge diversity was achieved with an interesting method for seeding: geotextile curtains were draped in the ocean for several weeks until fully colonized.

PACIFIC WHITE SHRIMP exhibited good growth rates on a diet of red wriggler worms and worm compost. They would hug the worms with their swimmerets and eat them like candy bars. To our amazement, they fed on worm compost as well. We theorize they were eating fungi or bacteria.

FLOATING PLANTS Spartina alterniflora was rafted in net pots set into rigid foam insulation. In addition to direct uptake and hosting microbial communities, the roots supported large numbers of tunicates.

Saltwater aquaculture eco-machine

Eelgrass ecosystems are my teacher. I am applying what I learn from them to design living technologies with comparable intelligence. My son, grandson, an associate, and I designed an eco-machine in the greenhouse attached to our house. It was designed to grow tropical shrimp and bay scallops and is a saltwater system composed of five interconnected clear-sided tanks that support a diversity of marine organisms. To obtain them, we submerged curtains in the ocean over the eelgrass beds. Within weeks they become colonized by marine creatures. We then removed the curtains from the sea, took them to the greenhouse, and placed them in tanks where they provided shelter and food for cultured species. So far, we have had success in growing tropical shrimp. The next challenge will be bay scallops. The illustration above delineates this small saltwater farm. The building of the system was simple. The assembling of the marine ecosystems was an on-the-water adventure for me.

Many natural eelgrass communities are dying or in decline and, with them, the bay scallops. The cause is pollution, coming especially from lawn chemicals and sewage. Consequently, the eelgrasses are being choked by excess algae growth and die. To prevent this, pollutants must be removed from the water, both at the source and in salt ponds and bays along the coast. We are designing and building floating restorer technologies to remove such nutrients as well as eco-machines to purify sewage on land. They are being adopted, albeit slowly, in several locations around the world. I wish the process were faster and that it was taking place in my own community. I don't want my greenhouse eelgrass ecology to become a memorial to an aquatic plant upon which so much depends.

9

A Plan to Heal Marine Bays and Salt Ponds along the Atlantic Coast of North America

Since being introduced to the ocean at twelve, I have been fascinated by it. I have worked as a professional oceanographer and have studied the behavior and communication of fishes. My studies have also led me to investigate the impact of pollution on the lives of a variety of marine creatures.

My heightened awareness of the dangers of pollution has affected me deeply. Thirty years ago, I decided to see if I could remove pollutants and nutrients from sewage, toxic wastes, and other contaminated bodies of water. As I have described in previous chapters, this effort led to the invention of a family of living technologies that I called eco-machines.

The massive oil spill from a damaged oil rig in the Gulf of Mexico during the summer of 2010 further woke me to the plight of the inshore oceans. The longstanding dead zone in the Gulf had been a constant reminder of the widespread problem of pollution caused by the leaking of excess nutrients into the environment. A 2010 global map of that dead zone and other threatened zones of the world was yet another catalyst. I realized that many of the heavily damaged zones are in the same areas as most of the great ocean nurseries. In the United States, Chesapeake Bay is a notable example of nutrient overloading and subsequent ecological decline. Coastal pollution is reducing more than local species of organisms. Oceanic food webs are at risk. The problem is worldwide and growing.

On Cape Cod, I also observed some areas that were changing rapidly. Eelgrass communities were dying back with a consequent reduction in bay scallops. Bay scallops, my favorite seafood, are culinary gems whose survival depends upon their symbiotic association with eelgrass communities. The causes for their decline in many of our bays, harbors, and salt ponds were easy to uncover. Excess nutrients were seeping into the coastal waters from the groundwater and from runoff from the rains and storms. Septic tanks and sewage treatment plants added to the problem and were also an important contributing factor. Such nutrients cause excessive algae blooms and die-offs as well as loss of water clarity. Normal biological communities are overwhelmed by the boom-and-bust cycles of the polluted zones.

Preventing ocean pollution has been widely recognized as an important goal for the last half century. People have become aware of the dangers of treating the oceans as garbage dumps. Conventional wisdom has focused on solving the problem at the source of the pollution, which is the right and proper thing to do. However, there is a serious omission to this approach that revolves around three issues. The first of these is legacy pollutants, meaning chemicals already in the environment that have been traveling through the ground toward the sea for decades or more. Rains constitute the second serious factor. In our area, it has been calculated that a significant portion of the nitrogen pollution in Buzzards Bay comes from polluted air in the rain.

The third issue is that of determining responsibility in both a legal and political sense. People tend to avoid spending extra money on fixing a problem that seems remote or distant. And whereas laws and regulations can ameliorate problems, they cannot eliminate them. Years ago, when operating an eco-machine to treat a modest percentage of Providence, Rhode Island's sewage, I saw the city's sewers flooded with toxic metals on holidays, like the Fourth of July and Christmas Eve, when no one was around to monitor offenders.

An alternative is to have a second line of environmental protection defense in the contaminated bodies of water themselves. Solutions are needed that are in place within our salt ponds, harbors, and estuaries. We have used this approach in freshwater bodies and the approach is applicable to inshore marine waters as well. In the early 1990s we built our first eco-technology called a pond restorer. It was installed in a highly polluted thirteen-acre pond with seasonal oxygen deprivation not unlike the dead zone in the Gulf of Mexico. The restorer changed all that. Circulating on average about thirty-five thousand gallons per day through nine internal treatment cells, the restorer revived the whole pond, waking it up and restoring its health. Over the years, we have improved the technology and adapted it to many environments. These have included sewage-laden canals in China and an award-winning salt pond restoration in Hawaii. At the largest scale, in Berlin, Maryland, one of our restorers treated 1.25 million gallons per day of slaughterhouse wastes contained in treatment lagoons. This is equivalent to the waste from a sizeable village or about four Olympic-sized swimming pools of waste. One result was significant reduction of nitrogen contamination in nearby Chesapeake Bay.

In a similar vein, we have a project for Dallas, Texas, on the drawing boards, that addresses parallel problems. The city is planning to create two

China restorer

new lakes that will flow into the Trinity River. The only source of incoming water will be 55 million gallons a day of treated sewage water from the city's treatment plant. As the lakes will be used for recreation, it is essential to remove the pollutants from the water before they are converted to algae blooms that would render the lakes unappealing and even toxic for boating, swimming, and fishing.

What is needed in Dallas is a technology to remove up to ten thousand pounds of nitrogen per day, in addition to four thousand pounds per day of phosphorus. Considering the astronomical costs of chemically and mechanically removing the nutrients, this represents a huge challenge. Our solution was to design a living technology that is also in a floating park. The park will double as a platform where the public can enjoy the lake and its clear waters. The design includes a series of large circles that houses the restorer eco-machines. The effect would be grand and beautiful. The circles would serve as wonderful platforms for swimming, kayaking, sailing, and picnicking.

The inshore waters of New England and the rest of the U.S. Eastern Seaboard present a greater challenge to on-site intervention as they are crowded with moorings and channel passages and space is very much at a premium. Additionally, there are only a modest number of locations where conventional floating restorers can be situated. Facilities like the proposed Dallas floating park in the marine environment would be wonderful and might be appropriate in larger open bays. In most places, however, an alternative solution to nutrient uptake must be found that is technically rugged to withstand hurricane-force winds and heavy seas.

In the town of Falmouth on Cape Cod where I live, there is an initiative for a major increase in the culture of mollusks, especially oysters but also mussels and clams, to improve the quality of the water. The town has granted ocean bottom space for shellfish culture. Oysters and mussels are very efficient at filtering particulate matter from seawater. This includes the microalgae that comprise a large percentage of their diet. In sufficient numbers, shellfish benefit water quality. If free from contamination, they can be a significant economic resource for the community. The twin goals of water quality improvement and economic benefit can thereby be met. Such a trend is likely to grow in the years ahead. As techniques for monitoring shellfish quality and contamination improve and become more readily available, the use of shellfish as a water quality improvement tool is likely to grow as well.

Yet another way to improve water quality along our coasts is with shoreside facilities. Constructed salt marsh wetlands and/or eco-machines could be built near the shore and shallow waters pumped through them for nutrient uptake. As their output would match the seasonal and day length needs of the systems, they could be powered with solar electric panels. The stumbling block with this approach is a general lack of space along the shore to set up these nutrient-removal technologies. It is unlikely that people in such a valuable and limited area would be willing to give up space on their land, even if there was legislation that would permitted it.

There is still another path to solving the problem. It occurred to me that boat moorings might be redesigned to serve double duty as micro-marine restorers. There are tens of thousands of moorings along the East Coast and most of them are located near the most heavily impacted zones. With my friend Matt Beam, I have been designing micro-marine restorers, in which everything is conceived in miniature. The technological challenge is to find

Red Brook Harbor, Cape Cod

ways they can be designed to circulate water in small spaces using only natural currents. Life-support surfaces must be designed for species from all of the kingdoms of life to reside within. Photosynthetic organisms, grazers, filterers, decomposers, and detritus eaters as well as predators must all be on board, literally. Fungi adapted to saltwater will also be needed.

The drawing below is a schematic of a micro-marine restorer adapted to moorings. I have not given any of the details, but it is possible to see the basic concept. The subunits are attached to a central boat mooring chain. From studying the colonization of surfaces in inshore marine waters, I have learned that systems like these quickly attract a diverse range of organisms on a variety of types of surfaces. In some mooring based-restorers, we will intentionally add organisms, especially bivalves or shellfish. All these organisms, in one way or another, will aid in removing excess nutrients and in some cases to convert nitrogen compounds to gases to be released into the air. Hopefully they will function in concert as a powerful tool for nutrient removal at mooring sites.

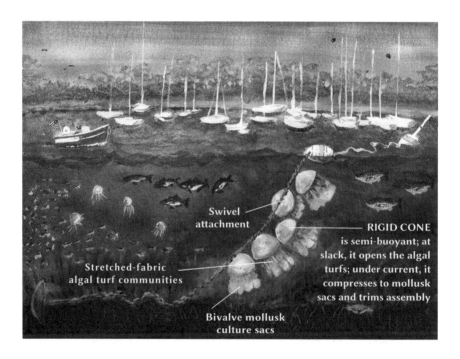

One hurdle for this plan to be implemented is their credibility as a treatment technology. It will be necessary to quantify how much excess nutrient each micro-marine restorer can remove. Each of them will be eighteen feet long and be composed of five subunits containing the biological elements. In Red Brook Harbor, which is our proposed test site, there are about five hundred moorings and, therefore, the potential for five hundred micro-restorers. To calibrate their total effectiveness, we will need to know how many pounds of nitrogen per year will be removed by each restorer. If, and this is hypothetical, each restorer removes twenty pounds of nitrogen a year, then, in all, ten thousand pounds of nitrogen would be removed. This would be a significant reduction in pollution levels.

To obtain such backup information, I am proposing that a nutrient removal test facility be installed along the shore. The facility I have designed is shown on the following page. It will be composed of ten five-foot diameter clear-sided tanks. They are connected like beads on a string to form the equivalent of a river. There will be five micro-restorers composed of two subunits for the biota inside each tank.

Seawater will be pumped from the harbor into the nutrient removal test facility, which holds about seven thousand gallons. New water entering the first tank will cause an equivalent amount from the last tank to flow back into the harbor. Flows can be varied to determine rates that optimize nutrient removal, which should be between twelve and twenty-four hours. The chemistry of the input seawater will be monitored and compared with that of the water leaving the system. The difference between the two sets of numbers will give us an estimate of the quantity of chemicals the test facility is removing. That number divided by the number of micro-restorers will give us the removal rates per restorer. The story is a little more complicated than this, but our goal would be to calculate the overall effectiveness of converting moorings to marine micro-restorers. If the numbers indicate that the technology can significantly improve the quality of inshore coastal waters, we have a case for their widespread adoption throughout New England, along the eastern seaboard, and eventually around the world.

Facility for testing the performance of mooring restorers

The cost for such a facility along with a one-year testing and evaluation program that would include oversight by a scientist will not be cheap. Compared to its potential benefits, however, the cost could prove to be a bargain. Prior to drawing up a detailed budget, I estimated that a test facility, including its fifty marine micro-restorers, will cost about $70,000 to build. This figure includes the fabrication of the prototype micro-restorers. The institutional operations, research, and scientific oversight would cost an estimated $120,000 for the first year. All told we are looking at roughly $250,000 to launch and test the concept.

Last summer the algae blooms in some of our protected waters were alarmingly extensive. Toxic blue-green algae, also known as cyanobacteria, appeared in our coastal ponds and produced cyanotoxins that can be dangerous and even deadly to animals and humans. Culturally we are at a critical turning point. It is imperative that we not only prevent new pollutants from infiltrating the waters, but that we begin to remove the chemicals causing the problems. Our proposed project described here shows how it might be done.

10

Caribbean Futures

It was one of those rare New England winter days. The breeze was slight and there was even a touch of warmth on my back as I pushed my skiff down to the water's edge. The sky was filtered yellow and looked somewhat otherworldly. The town seemed empty of people and cars. Looking down the harbor, I saw that there was only one sailboat still at a mooring. The rest of the fleet has been hauled and taken out of the water. I could make out two people in the cockpit of the lone sailboat. The larger of the two was setting sail. It was a perfect January day for a sail.

The boat cast off and sailed towards me. An exquisitely shaped double-ended ketch, it was so much part of this watery element, ghosting along silently and quickly in the gentle breeze. A lovely blonde girl of seven with her father at the tiller waved as they passed. I watched them sail up to a dock at the other end of the harbor, pick up a passenger, a tall blonde woman, who I guessed was the mother. She grabbed the stay and jumped aboard. I watched them disappear around a point to the east, and I wondered why many people haul their boats so early and launch them so late. I knew it is prudent to do so to avoid winter weather damage, and I was aware there are insurance issues, but these short days with the sun low in the sky could be wonderful on the water. Maybe that is why I too was willing to take the risk of keeping a boat available for days like this.

A few days later, I packed my bags and headed south to the Caribbean. A small team of colleagues had been hired to create coastal development ideas for Vieques, a small island off Puerto Rico. We drove through the night in snow and freezing rain and arrived at Kennedy Airport early in the morning.

By mid-afternoon we were in San Juan and about to hop on a smaller plane to Vieques. When we stepped out of the plane, the enveloping tropical warmth was a great tonic after the long day. I never tire of the first smells and sounds when landing in a new environment. It had just rained and the frogs, called "cochi" because of the sound of their call, were starting to warm up as the sun went down.

Vieques is located just a few miles to the east of Puerto Rico. It is about eighteen miles long and runs in an east-to-west direction, and is about four to five miles wide. The center of the island then had two main settlements where most people lived: Isabella Segunda on the north shore and Esperanza on the south. Esperanza was nearer the beaches and has several inns and guesthouses as well as waterfront restaurants and shops. Recent tourist developments throughout the Caribbean had passed this area by. The explanation, in part, for the out-of-time feel to this place was the presence of the United States Navy. The navy controlled over half of the land area, using the western end for munitions storage and the eastern for maneuvers.

There were wild horses everywhere called "*paso fino*" after their high stepping gait. There was not much agriculture or gardening and most of the food was brought in from Puerto Rico. Water came from a seabed pipeline that also originates in Puerto Rico. There did seem to be an active fishery as I counted over thirty boats at the harbor. These vessels were long, narrow, open skiffs with a lot of flare to their forward hulls, presumably for protection from spray. Most were powered by twin seventy-five or eighty horsepower outboards. The commercial fishermen appeared to be important members of the community.

We were told unemployment on the island is very high, unexpectedly so. Vieques was then hard-hit during Hurricane Hugo in 1989 and most recently it was devastated by Hurricane Maria in 2017; with frequent stormy weather it was difficult to build the economy. I found myself exploring the wreckage of a resort that was never rebuilt. Broken windows, tattered mattresses, upside-down toilets, and smashed cabinets lay strewn throughout the rooms. A large building, still without its roof, had trees and vines growing out of the cracks in the cement floor. Termite nests and spiny leguminous trees dominated the landscape. I suspected that many a spirit was broken by the hurricane.

Among the remarkable attributes of Vieques then were the exquisite beaches along the southern shore. There were miles of sandy bays and coves protected from the surf. The movie of William Golding's *Lord of the Flies* was filmed here. One of the film crew was our botanical guide on this trip. Mangroves grew in and around some of the brackish pools, and ringed a most unusual and protected water called Phosphorescent Bay.

Phosphorescent Bay was an ecological wonder. At night, electrically powered catamarans took people on outings on the water, which burst into a blaze of light as the hulls of the boats plied the lagoon. Each fish glowed with its own distinct shape. On our trip we watched a luminescent manta ray rise off the bottom and swim away, its tail and fins making it look like a space ship in a distant galaxy. Diving into the water in the dark was an out-of-this-world experience. Thousands of tiny lights flowed over my skin, mirroring the countless stars in the sky above and reflected in the water below. Glitter gathered on the chests of the hairiest males. With a face mask on, the experience became psychedelic. It was hard to leave the lure of the light, but even in the Caribbean I eventually got cold underwater.

The stars in Phosphorescent Bay are tiny unicellular marine plankton called dinoflagellates. At night, they give off light when they are agitated. This is caused by a chemical they produce that is, quite appropriately, named "luciferin." Luciferin is broken apart by an enzyme, luciferase, which is only produced at night. What makes Phosphorescent Bay such an incredible nursery for this vast population of dinoflagellates is not fully understood. Circulation and nutrients are certainly factors. It is suspected that dinoflagellates are sensitive to pollution, as many formerly well-known luminescent bays have lost their luminescence and no longer glow brightly. However, the bay on Vieques is protected and there are efforts to keep motorized vessels away. Gasoline or diesel spills could be disastrous to the little creatures there.

Several miles to the southwest there is another bay, which is adjacent to a popular holiday campground and is perfect for swimming. One of our tasks was to figure out why the beach at the western end of the bay is eroding. There were exposed coral outcrops and swimming had become precarious. A decade earlier there had been wide beaches in the same location. It was thought that the erosion process was natural and was caused by a current sweeping through the bay. Expensive schemes had been proposed to halt the gradual destruction of the beach: one idea was to build underwater structures offshore that will change the current's direction. Another scheme under consideration involved filling the bay with acres of a plastic "kelp" to dampen the flow. The idea was that as the water slows it would deposit sediments and thus build up the beaches.

I was suspicious of both approaches for two reasons. The first was cost. Many dollars could be spent without any guarantees of success. The second reason came from snorkeling in the area. I discovered healthy beds of seagrasses in some places, abundant patches of turtle grass, manatee grass, and shoal grass. Calcareous algae, which have calcium carbonate in their tissues, also grow amongst the grasses. Beach sands are primarily composed of particles from decomposing calcareous algae. They are literally beach makers.

I then surveyed several healthy beaches and returned to the eroding beach for a second look. There was a difference between the eroding beach and the healthy ones. The healthy beaches were ringed by dense groves of trees. Palms, legumes, and beach grapes predominated, and beach convolvulus, a spreading sand-binding plant, literally carpeted the ground above the spring high tide mark. The vegetation that protected the beaches was intact.

At the eroding beach there were no trees, except for a few recently planted palms and some mahogany trees that were well back from the water. The beach had been turned into a lawn for campers. The banks had been cut into by the large numbers of people who visit during festivals. Beach convolvulus did not carpet the upper beach. The natural regeneration of the vegetation after tropical storms and hurricanes had been discouraged by park managers. The consequence was a sadly eroding beach. The good news was that there was a solution and one did not need to own a bank to solve the problem.

Towards the late afternoon, we walked along the beach to the town of Esperanza. Its entrance was marked by a large cement dock that jutted out into the harbor. There were remnants of an old railcar loading system that had been used in the days when sugarcane was king here. The dock was last used for shipping pineapples, but those days were long gone too. It was on the dock that we began to plan for an ecologically based agriculture for the island. The planning took months to complete, but it was based on our

Beach convolvulus

experience with intensive food production at the New Alchemy Institute on Cape Cod. The plan was to be diverse, intensive, and balanced between trees, row crops, and animals. It also had a large element of aquaculture and marine aquaculture recommendations.

Leaving the dock, I spotted the wreck of an old Tortuga sloop pulled up on the shore. Despite its condition, it still looked beautiful and timeless. I was reminded of a Winslow Homer painting of a dismasted West Indies sloop struggling along, perhaps fatally, with its lone sailor, bereft of hope.

Leaning against the sloop, we discussed an alternative future for life on the waters of the Caribbean, one that would include fishing, ocean transport, and recreation based upon island resources. A number of years ago, I had been involved with the brilliant boat designer, Dick Newick, in developing high-performance sail-powered working craft for coastal people in the Caribbean and South America. Our prototype was built on Martha's Vineyard in New England. It was a thirty-three-foot-long sea-going trimaran. We called it the *Ocean Pickup* because of its utilitarian role. It was built of cold-molded composite materials and used fast-growing and lightweight woods.

The original *Ocean Pickup* was intended for use by local fishermen in demonstration projects. We sailed it, via Bermuda, to Guyana in South America. It worked successfully in Guyana as well as along the Caribbean and Pacific coasts of Costa Rica in Central America. During those years, we experimented with making composite materials using wood from local fast-growing trees and imported epoxy materials. Our lightweight as well as very strong composite materials, proved a success. We knew that it was possible to build high-performance sail-assisted working vessels that could meet the needs of local communities. Also, we proved they could compete successfully and economically in several tropical fisheries.

Back on Vieques, the conversation turned to planting horse pastures and former sugarcane fields with fast-growing shrubs and trees. In a few years, such trees could be used to build boats our new way. The bulk of the materials could be grown locally, and modern vacuum-bagging techniques used for fabricating composite materials could readily translate into an island setting.

Ocean Pickup

We began to think about building whole fleets along the shore. They would include fishing craft and small, sport-sailing multihulls as well as eco-tourist excursion and diving boats. Later in town over beers, fried snapper, and cooked plantains, we began to spin these ideas out further. We knew from our wastewater treatment projects in the Caribbean that transporting goods and people could be difficult and expensive. It was not always possible to fit our needs to local forms of shipping and transportation. Why not, we rationalized, design and build a new generation of inter-island travel and cargo vessels? We could calibrate our lives to those of the trade winds rather than the dictates of air travel or high-speed ferries. The idea of the voyage being as meaningful and pleasurable as the destination seemed too good not to explore.

I contacted the legendary British yacht designer Nigel Irens who is famous for the speed of his racing boats. Both his sailing and motor-powered multihulls held records for the fastest voyages around the world. He was also deeply rooted in the traditions of working watercraft. For Ocean Arks International, our nonprofit, Nigel had designed the long and narrow light cargo vessel shown here for Vieques. It was a motor-assisted sailing vessel intended to be fast, fuel-efficient, and cost-effective. Its rig doubled as cargo handling gear.

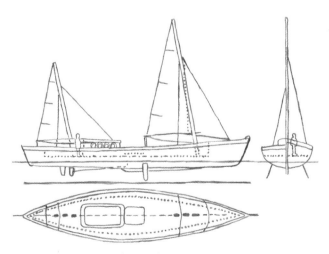

Nigel Irens's light cargo vessel

The relic Tortuga sloop I saw on the beach echoed a great tradition of living and working on the sea. It further served as a reminder that throughout history the wind has been a guiding force, determining the speed at which we explored the world. I yearned to live at Aeolian speeds—speeds dictated by the wind. Maybe there in the islands these ideas can take root. Maybe we should call the first of the boats *Convolvulus* after the modest plant that protects the shore at the edge of the sea.

Since the 2017 hurricane I have joined the New York architect and landowner on Vieques, Roberto Brambilla to help address the island's needs. He is organizing a nonprofit foundation and resources to begin work on the creation of an infrastructure for the island designed to cope with tropical storms and rising sea levels, and provide self-sufficiency in water, energy, and food. His plan includes an educational effort to train the islanders in the technical, construction, and operational aspects of the emerging ecological infrastructure. His dream is that Vieques become the jewel in the crown of the Caribbean.

11

Solar- and Wind-Powered Workboats

Ocean Arks International, the nonprofit organization I founded together with my wife, has a long tradition of exploring ecological ideas related to the sea. For over thirty-five years we have been involved in ocean transportation concepts. With the naval architect Phil Bolger, we started to develop an Ocean Ark, an ecological hope ship, named the *Margaret Mead* after the anthropologist who was our friend and mentor. It was to be a sailing ship, the size of clipper ships of the nineteenth century, capable of delivering biological support materials, including food and tree crops, to impoverished coastal regions around the world.

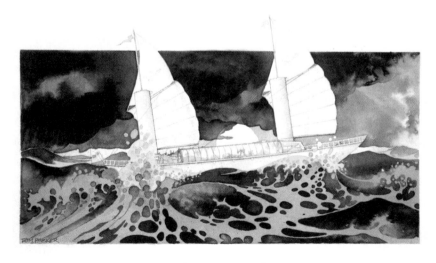

The *Margaret Mead*

Prototype Vessel, the *Nancy Jack*

We built a one-fifth scale, fifty-five-foot prototype of the Ocean Ark we called the *Nancy Jack*. It was designed to test several unique concepts proposed for the full-scale vessel, including a somewhat radical rig with wing masts, water ballast, twin centerboards, and bow and stern steering. From this we learned so much about future directions for commercial sailing and gained a great deal of respect for Phil Bolger's innovative ideas.

The Ocean Ark project was followed by a more modest approach to sail-powered workboats. I was approached by the Canadian International Development Agency and asked if I could develop the marine equivalent of a pickup truck, in the form of a working vessel that would be primarily powered by the wind. This Ocean Pickup, which I mentioned in the last chapter in relation to the needs of Vieques, would be developed for coastal communities that lacked, or were restricted, in their access to fossil fuels. We had in mind countries like Guyana in South America with soft currency that made gas or diesel purchase almost impossible for poor fishermen.

We turned to the brilliant multihull designer Dick Newick, who was revolutionizing yacht design with his race-winning sailing trimarans and

Guyanese fishing vessels

proas. Proas are Polynesian sailboat designs with two hulls, one larger than the other, fast boats best only handled by very experienced sailors. He had a well-deserved reputation as a man concerned with the needs of poor coastal communities. Dick designed and helped build the first Ocean Pickup. We named it the *Edith Muma,* after a great lady who had supported our efforts. The *Edith Muma* became known as the one-and-a-half-ton Ocean Pickup. The number was based upon its cargo-carrying capacity.

Dick Newick went on to design both larger and smaller Ocean Pickups based upon the varying needs of fishermen. My son and I, along with a friend, an expert sailor, then set out and sailed the thirty-three-foot-long *Edith Muma* from New England to Guyana. Once there, we successfully developed less harmful methods of capturing and transporting marine resources. The *Edith Muma* subsequently worked in Costa Rica on both the Caribbean and Pacific coasts. After seven very active years, Ocean Arks International donated the boat to the University of Costa Rica for its marine research.

Ocean Pickup on the Caribbean Coast of Costa Rica

Since 2003, I have been working with green developers in the Caribbean with their food, energy, water, and waste problems. Our most recent project was a recently completed wastewater treatment and water reuse eco-machine for a new development by Sir Richard Branson in the British Virgin Islands.

We have ecologically engineered shoreline restoration schemes, island agriculture and aquaculture systems, as well as eco-machines for total water management and reuse. The materials and supplies for these projects come at a cost to the green image of the projects and the environment. Currently inefficient fossil-fuel-powered boats provide supplies and fuels to the islands. Perishables like fresh produce and meats are usually flown in, carried by small and relatively inefficient aircraft. We realized that it is transportation that stands between the islands and a carbon-neutral future.

Concern for the inconsistency and contradictions between developing green communities and their huge transportation needs initially gave me pause. The way forward in my view was to develop a new kind of transportation and cargo craft employing design based upon net-zero solutions. Our proposed vessel would be a hybrid, powered with advanced wind propulsion and solar electric engines. It would be the "Tesla" of the sea.

In 2016 I looked for a marine architect with whom to collaborate. I needed someone creative who had designed boats of varying types that solved problems. He or she had to have an interest in helping the environment and finding local solutions for materials and construction. My plan was to have the boats built in the islands, but have very advanced designs, as we had done with the Ocean Pickup years before.

Sadly, Dick Newick had passed away in 2013 since working with me on the Ocean Pickup, so I had to look for another naval architect with the requisite skills for the task. Once I discovered Laurie McGowan, a marine designer from Nova Scotia, I knew I had found the person for the job. He was developing biplane-rigs for twin-hulled catamarans that were capable of dramatically increasing their power. What convinced me of his ability for this task was the intelligence and creativity he and his associate Michael Schacht had put into the design of a sailboat called *Chimaera*, where they solved difficult and normally competing design and technological issues.

It did not take us long to come up with a general concept. I wanted a hybrid solar/wind-powered boat that could keep to a schedule in a range of weather conditions. The prototype would have to be a water taxi and a light cargo carrier. I decided to name the boat *Pelicano*, the Spanish name for pelican, one of my favorite seabirds. It was not long before Laurie McGowan came back with the concept illustrated on the next page.

Pelicano is a catamaran of forty-three feet in length and twenty-two feet in width with solar electric and advanced wind power, a hybrid vessel powered by the sun and the wind. It is intended to be a game changer on many fronts. It is relatively fast with a working speed of ten knots, able to keep to a marine schedule. It is also wheelchair-accessible and capable of protecting cargo and passengers in inclement weather. It is seaworthy and can land on protected beaches as well as conventional docks.

It is designed from the very outset to be able to be produced in local boat-building shops, following testing of the prototype. My prior boats, including the Ocean Pickup and the prototype Ocean Ark, were all built and tested first in New England. We have not yet selected a builder for *Pelicano* so I do not know where it will be built. It is my hope that the *Pelicano* series of vessels will help reestablish boat-building skills and employment wherever they are utilized. We expect the first *Pelicano* will be used in the Caribbean.

I was very interested in *Pelicano* being a water taxi and a light cargo carrier, but Laurie McGowan came up with a flexible design that could be adapted for fishing, diving, and excursion boats. It could be scaled up and down in size. Another advantage of the design is that its twin hulls balancing the boat will allow it to go right up onto beaches in fair weather.

Pelicano will be a harbinger of a carbon-neutral era on the sea. It may well be the first water taxi and light cargo carrier that can keep to a schedule powered only by renewable energy.

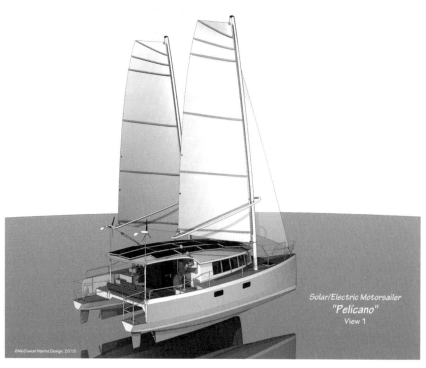

Solar/electric motor sailer *Pelicano*, view 1

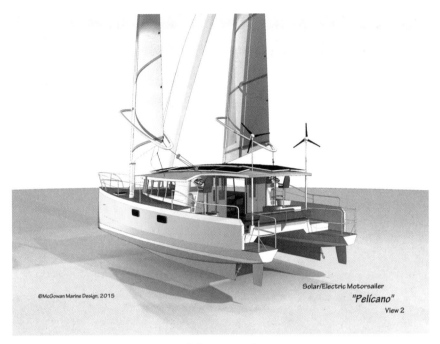

Pelicano, view 2

The following are *Pelicano*'s technical specifications:

Length overall: 43′

Width: 22′ 3″

Draft or depth below the water: 2′ 9″

Displacement: 14,500–22,500 lbs.

Working sail area: 800 sq. ft.

Electric motors: LEMCO Swordfish 26 V Twin (26kW × 2)

Wind generators: 2 × Air Dolphin 1kW

Solar Panels: 10 Kyocera KD200-60 + 8 Kyocera KD 135 for 3.58kW total

It can carry thirty-two passengers and require a crew of three.

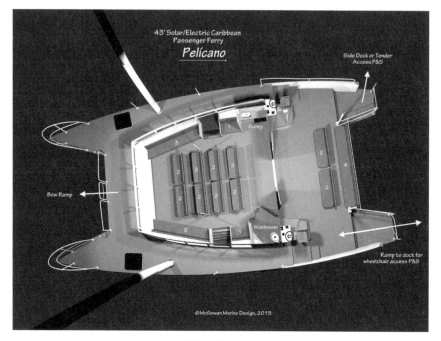

Pelicano's interior

I estimate that the prototype will cost something just under $1 million. Planned regional production lines will reduce these costs significantly in future vessels. However, it is my belief that even the prototype will quickly pay for itself in various marine services. *Pelicano* is a harbinger of a carbon-neutral age.

12

Cleaning Up an Ongoing Oil Spill with Eco-Machines

Fifteen years prior to writing this book, I met a remarkable man by the name of Eugene Bernat, known as Gino. Gino had acquired twelve acres of land that had been an old mill site situated along the Blackstone River Corridor in the town of Grafton in central Massachusetts. The area was part of the birthplace of the American Industrial Revolution. The river flows from the town of Worcester, Massachusetts, to Providence, Rhode Island. At the beginning of the industrial era the river was dammed and river water powered the mills that lined its banks. Parallel to the river, canals were built to transport goods and materials the length of the river. Gino's land had once been the site of the Fisherville Mill, which had burned down in the latter part of the twentieth century in what the local fire chief described as a horrendous blaze.

The remains of the ruins, mostly rubble, occupied a remarkable piece of land. To the north, the river widened into a large, shallow, slow-moving lakelike area. It was surrounded by trees and marsh, bounded on the east by a dam, and it had a feeling of wildness. The river dropped considerably at the dam, wrapping itself around the site before heading south on its way to the sea.

Gino had been dreaming of a twenty-first century village for the site. He wanted the project to be green and to have a very low environmental impact. Gino made his living converting waste materials to valuable new products and has found waste-derived substitutes for petroleum-based asphalt used for paving. He also had a long-time interest in the creative management of organic wastes, including large-scale production of compost for agriculture. Knowing the former Blackstone River mills had given birth to the first

Industrial Revolution, Gino wanted his project to be a catalyst for a second environmental revolution with technologies and techniques for managing resources adapted to a new ecological age.

When I toured the site with Gino and listened to his vision, I was impressed with his account of the hurdles ahead. For all his confidence, they seemed like more than enough challenges for any one man. The site was contaminated with toxic materials. A contaminant of real concern was number six or Bunker C oil, which had been stored in tanks that had subsequently ruptured. Bunker C oil is a toxic, tar-like residual material or sludge from the manufacture of petroleum products. It is used as a fuel in ships and in electrical power plants. Old storage tanks buried in the ground were breaking apart and the leaking oil was contaminating the groundwater, seeping into the canals and ultimately into the Blackstone River. Gino hoped our living technologies in the form of eco-machines could decontaminate the heavy oil that prevented more positive development.

We explained that we had never treated Bunker C oil but had developed eco-machines that broke down such chemicals as DDT and other noxious pesticides, which are hard to decontaminate, but we welcomed a chance to test our technologies on such heavy oils. After several years of political and financial negotiations, Gino located the funds to build a pilot facility to test whether an eco-machine could treat Bunker C oil. We subsequently did so in a greenhouse owned by the Woods Hole Oceanographic Institution overlooking Nantucket Sound and Martha's Vineyard.

We built our pilot eco-machine in the fall of 2006. It was made up of two parallel treatment systems. Each had four ecological components through which the contaminated water and oily sediments flowed. The total volume of the two systems was approximately four hundred gallons. The first three components were housed in clear-sided tanks that allowed sunlight to penetrate. Each tank contained a different ecology. The first housed algae communities, which we grew on screens. The second tank had specially designed rafts that supported marsh plants on their surface. Their roots grew deep into the water column and were colonized by a diversity of microbial life. The third tank was an open water tank that supported microalgae, plankton, and fish. The fourth and final component was housed in dark plastic chambers containing fungi in the form of mushrooms. The fine networks of mycelia produce enzymes known to degrade many compounds, including, we hoped, number six or Bunker C oil. The fungal component acted as a trickling filter. Standing water was not allowed to accumulate.

After the fourth stage the liquid was recycled back to the beginning. It ran on a continuous loop. We had collected the contained organisms from half a dozen aquatic environments ranging from freshwater streams to salt marshes. We introduced thousands of species, which quickly began to self-select, self-design, and self-organize into unique ecological systems adapted to the waste stream. We completed the inoculation period in December of 2006 and operated the pilot eco-machine from January until April 2007.

The system proved effective in treating the heavy oil. We made chemical measurements of Total Organic Carbon (TOC) and Total Petroleum Hydrocarbon (TPH) from the canal sediments, from the water itself, and from the eco-machine. In mid-December, we extracted ten pounds of sediments from the canal and split them equally between the eco-machine's two treatment systems. From January through March we added close to six hundred gallons of canal water. Despite our continuing addition of contaminated canal water, by the beginning of April over 90 percent of the Total Petroleum Hydrocarbons (TPH) had been removed from the water. The volume of the oily sediments had been reduced 57 percent in one of the treatment lines and 89 percent in the other. Just under 50 percent or between 40 percent and 56 percent were reduced in the remaining sediments' TPH. The pouch snails *(Physa gyrina)* that we had introduced were eating the oily sediments attached to the walls of the tanks.

By the time the experiment ended in early April of 2007, we were beginning to feel confident that ecologically engineered systems that employed representative species of all the kingdoms of life could tackle the decontamination of petroleum hydrocarbons, including such heavy oils as Bunker C. We then felt ready to work directly on the canal, but it was four years later before we finally got the chance to try.

With support from the U.S. Environmental Protection Agency, the town of Grafton initiated several environmental cleanup efforts for the Fisherville Mill site and the adjacent canal. The canal was dredged and contaminated spoils were trucked offsite. In addition, an attempt was made to inject butane into the groundwater under the site to make the heavy oils more susceptible to biodegradation. The goal of both projects was to reduce the overall contamination loads in the area. It was a prudent move on the part of the town.

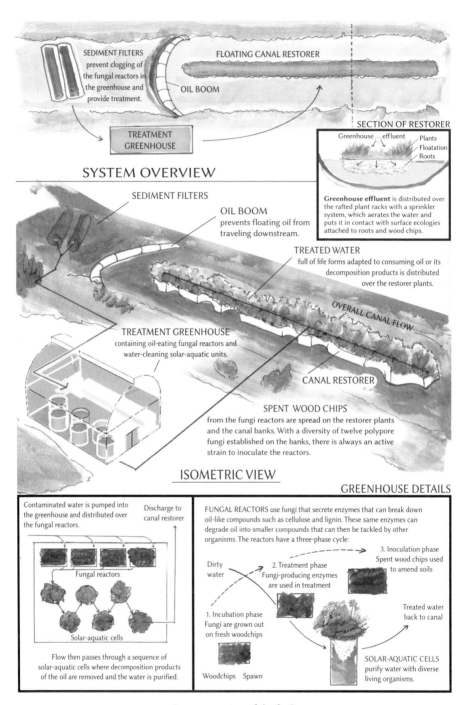

System overview of the facility

Our part came later. Gino had donated land adjacent to the canal to the town of Grafton, where a beautiful public park had been created. It quickly became a favorite spot of local people who used it for picnicking, sports, and public events. The site for our eco-machine and canal restorer technologies was along the western edge of the park, which was a great location offering public access and visibility for our work.

My son Jonathan led the design team. For the Grafton canal cleanup, we decided to create a hybrid technology specific to the site. On the canal, we installed a floating restorer of the type we had developed for the Baima Canal in Fuzhou, China, and for a slaughterhouse waste treatment lagoon in Berlin, Maryland. The restorer's role in Grafton involved circulating contaminated canal water in and around the massive root complexes of the higher plants being grown within. Its purpose was to provide habitats for beneficial organisms that would improve water quality.

The canal and the facility

Oil slick

We placed the second technology on the bottom of the canal. It included a bio-filter through which sediments and canal water circulated before flowing to the eco-machine and the floating restorer. We originally developed this bottom technology for an aquaculture facility at the Four Seasons Resort in Kona, Hawaii. Thanks to the restorer, the resort's saltwater pond now supports large populations of oysters, shrimp, and marine fishes for the resort's kitchens. The main purpose of the bottom bio-filter on the canal in Grafton was to host biological activity in a low-to-zero oxygen environment and to convert nitrates in the water to harmless nitrogen gas. The bio-filter also degraded organic compounds in the sediments and adjacent waters.

The third technology for Grafton was an eco-machine situated on the banks of the canal. It was housed in a greenhouse and included boxes for mushroom cultivation and a series of clear-sided tanks through which canal water was pumped from the bio-filter and then made to trickle through the dark, enclosed cells containing fungi with rapidly growing mycelial networks. After the fungal system, the canal water flowed into a series of translucent tanks that housed complex solar-based ecosystems.

Cleaning Up an Ongoing Oil Spill with Eco-Machines

Greenhouse housing the eco-machine

The eco-machine with the fungi systems in the black cells on the right

The overall purpose of the eco-machine was to provide large numbers of beneficial organisms to the canal on a year-round basis. It functions as an ecological incubator providing a sufficient density of life forms from the various kingdoms of life to digest the oils and transform the ecology of the canal to a healthier state. Water from the Grafton eco-machine flowed back to the restorer zone in the canal. The concept of an ecological incubator was new and quite radical, but its potential for water quality improvement was very real.

The system's configuration is illustrated on page 106. The facility was inoculated and started up on the first of June 2012 when we began circulating canal water through the system. On June 14, which was Flag Day, there was an opening ceremony complete with a raising of the flag, speeches by local politicians and town officials, and a brass band. The New England Regional Administrator of the U.S. Environmental Protection Agency and the Commissioner of the Massachusetts Department of Environmental Protection were also in attendance. They indicated interest and excitement during their tour of the facility. By then the water in the eco-machine had been completely replaced by water from the canal. We were pleased with opening day and delighted by the beauty of the eco-machine and the floating restorer on the canal.

The following day we collected water and sediment samples from the canal and the eco-machine. They were taken to a laboratory at Brown University in Providence that specializes in the measurement of petroleum hydrocarbons. It

Floating restorer located mid-canal

was over four weeks before we got the first results. They were promising. The level of petroleum hydrocarbons in the water above the bio-filter, being the first step in the cycle, was 42,672 nanograms per liter (ng/l). A nanogram is a very small unit of measure, a billionth of a gram, but it is the conventional unit of measure for studying petroleum hydrocarbons in the environment. What is important here is the percent reduction of the oils. By the end of the eco-machine process, the levels had dropped to 5,385 n/l, representing a reduction of 87 percent.

This is quite remarkable considering the short period of several weeks during which the system had been running prior to taking the samples. Also, there were interesting results from both the upper end of the canal near the restorer and the furthest downstream sampling point. The highest number was downstream well below the restorer, the bio-filter, and the eco-machine. The meaning of this is not yet clear. One explanation is that the restorer had begun to clean up the canal in its upper reaches. At that moment, it was too early to come to any conclusion.

A second set of oil samples was taken on July 13 and again sent to Brown for analysis. What we now know is that the oil contamination levels in the canal water increased quite dramatically. Heat may have played a role in this. During the hot days of early July, a larger than normal sheen of oil was observed entering the canal. However, the oil levels leaving the greenhouse remained low at 7,851 ng/l. This represented a 99 percent reduction in petroleum hydrocarbons from the canal water.

The U.S. Environmental Protection Agency funding was not renewed the following year due to federal financial cutbacks to projects in New England. A consortium of universities including Brown, Clark, Worcester Polytech, and Tufts has continued to study the progress of the eco-machine on the canal. The facility has morphed into the Living Systems Laboratory under the dedicated leadership of Gino and his son, Nick. The ecological treatment system has performed remarkably well since the outset, having removed over 90 percent of the oil from the water that has passed through the greenhouse. Not only is it proving effective for oil treatment, it is also functioning as a biological incubator for a wide diversity of organisms that are proving beneficial for the rest of the canal.

The exciting news for me is that the canal is really coming back to life. The water in the canal looks so much better, despite the fact that number 6 oil continues to contaminate it. Frogs, normally very sensitive to petroleum pollution, are breeding and fish are becoming common. The canal does not have the stench of oil that it used to. The downstream effect is spreading.

Sadly, we no longer have the resources to study the details of the transformation in the canal in a measurable scientific way. The only instrument of measurement I can use are my own eyes. There is a vibrancy in the water and on the water, and I can see that the living technology continues to support the ecosystem in the canal, proving its cost-effectiveness year in and year out.

13

Ocean Restorers: Ecological Hope Ships for Marine Pollution Reduction

Beginning on April 20, 2010, the images of gushing oil, fires, and plumes of smoke over the Gulf of Mexico near the mouth of the Mississippi were ghastly to behold. The oil-soaked animals in distress and dying were excruciatingly painful to watch. And below the surface, I could only imagine whole oceanic food chains being torn apart unseen. The BP (British Petroleum) Deepwater Horizon disaster poured 4.9 million barrels of oil equivalent into the sea. In the end up to 68,000 sqaure miles were affected. Eight years later the area has still not recovered. The disaster was an affront to the very fabric of the sea and a miserable commentary on how our society conducts its business.

The great question for our time is, can we heal that which we have set asunder? Can a culture of Earth and ocean stewardship be created in the shadow of a global culture that consumes and destroys the very life upon which it depends?

I think so, at least in theory.

Watching the flames and the smoke rising off the Gulf of Mexico I began to envision a plan to restore the seas and heal the polluted coastal regions around the world. The plan I am about to describe may be quixotic, but it is a plan based upon real and tangible experience. The difference this time is the scale of the endeavor, as this plan is several orders of magnitude larger than anything we have attempted before.

Twenty-five years ago, as I recounted in chapter 7, I set out to resuscitate Flax Pond, a thirteen-acre pond in the town of Harwich on Cape Cod. It was badly polluted. Each year an adjacent landfill spilled over 20 million gallons of pollution into the pond. Like the infamous dead zone in the Gulf of Mexico, the bottom of the pond lacked oxygen. Visible animal life, known as benthos, were absent from most the bottom. The pond was "comatose" and the waters above was contaminated with hazardous chemicals and heavy metals. It had been closed to fishing and recreation.

On the most heavily impacted eastern end of the pond, we built our first restorer, a living technology designed to clean up the pond, a raft-like structure thirty-five feet long and just over twenty feet wide. A solar panel and a small windmill produced the electricity that ran its pump. The pump circulated thirty-five thousand gallons a day of water from just off the bottom through nine ecologically engineered cells that made up the restorer. What then happened to Flax Pond was quite extraordinary. Within weeks it began to "wake up" and the chemistry of the pond began to change. The restorer dramatically reduced levels of the chemical known as ammonia, which is toxic to aquatic animal life. Within months life began to return to the pond. The accumulated sediments on the bottom were digested and their depth reduced by almost two feet over the next several years.

Most remarkable was the fact that the beneficial effect of the restorer spread from the eastern end of the pond to the whole of the pond and its more than 20 million gallons of water. The restorer seemed to function as a catalyst and alter the whole system in beneficial ways. We were never able to fully explain why this was so. The Flax Pond restorer ran for ten years before it was dismantled due to old age, with the pond being reopened to the public for recreation and fishing.

Years later we built another restorer in Berlin, Maryland. It was placed in a large lagoon that received 1.25 million gallons a day of high-strength organic waste from a slaughterhouse that processed 1 million chickens each week. This restorer was much more technologically robust and advanced than the Flax Pond system, but the same ecological design principles were at work. The volume treated was thirty-five times greater. The concentration of the waste was even stronger. The floating technology employed in its design twenty-five thousand higher plants, including trees and shrubs. It worked. It saved the company more than 70 percent on its electrical bills for waste treatment and brought the processing plant into compliance with local discharge regulations.

Perhaps our most dramatic and well-known restorer was built in China on the Baima Canal in the city of Fuzhou. The canal was a fetid and odoriferous place as it received raw sewage from the adjacent large buildings, including skyscrapers.

Sewage-laden water flowed down the canal and was treated by our attractive and botanically beautiful restorer technology. The restorer met Chinese EPA performance standards for water quality. It reduced odors dramatically. Fish that were introduced into the canal thrived with the restorer's environmental support.

We also had experience using restorers to treat saltwater. As I mentioned in the previous chapter, two restorers are on a salt pond at the Four Seasons Resort in Kona, Hawaii. The salt pond is used to grow hundreds of thousands of shrimp, oysters, and native fish for the chefs at the resort. Prior to the installation of the restorers, the pond water was replenished by pumping in new saltwater at a cost of $10,000 a month. With the help of the restorers, the electrical bill is now just $400 a month. The water quality is excellent.

For years, I have been planning technological ways to protect our salt ponds and coastal waters in New England. The first-generation ocean restorers to come from our drawing boards were long ribbonlike floating structures moored to the bottom of the ocean. Our plan was to have the tidal currents move vast volumes of water past ecological communities attached to artificial materials suspended from the restorer platforms above. These communities would filter and produce oxygen for the water as well as contribute beneficial organisms to aid in the purification cycles.

Then the Gulf oil spill came along in 2010. At the time I was teaching ecological design at the University of Vermont, and my students and I talked about solutions to this devastating crisis. What the Gulf needed were ocean restorers that were not anchored, but would be propelled through the water in order to increase dramatically the amount of pollution they might treat. Our solution to the problem of propulsion was to envision an ocean restorer that was a ship, which might move through the water with its ecological

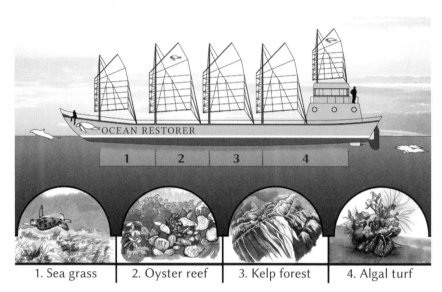

Ocean Restorer: **the original concept**

treatment technologies suspended between two hulls. The vessel would need to be a catamaran and have two hulls.

I planned the ocean restorer to be sail- and solar-powered. The original concept vessel had five sails on each hull. The rig is very much like a traditional Chinese junk rig. There are many reasons for this design. A number of yeas ago I sailed on a modern junk-rigged vessel off the coast of England and was mightily impressed with the ease with which it could be handled. First, the sails do not flap, as they are fully battened. Secondly, they are mounted so that they can swing freely and spill the wind if necessary. Thirdly, they do not require heavy machinery to raise the sails or manage their trim. But most importantly, the lack of flapping and the balanced rig would allow the sails to be made from new materials that would double as photovoltaic or solar cells. In this original concept design, each sail is both a propulsion device and a solar-electric-generating station, making the vessel a hybrid wind power/electrical generation device.

The prototype *Ocean Restorer* catamaran, at 130 feet in length, would be capable of sailing at up to 20 knots through the water without its ecological technologies submerged into the water. However, when lowered into the water with these eco-technologies between the hulls, we expect it will achieve a forward motion speed at a much slower 3 knots. The relatively slow speed of the restorer will be due to the drag or resistance of the treatment technologies suspended in the water. If we can achieve an operational speed of 3 knots, then the restorer will sweep and treat millions of gallons a day. It will be possible to raise and lower the eco-technologies between the hulls, and in the raised position seawater will be sprayed over the organisms to keep them well and thriving.

The most difficult design challenge we faced originally was not the solar/wind rig or the catamaran design, which require straightforward naval architecture and marine engineering. The great challenge will be the design and ecological engineering of the treatment technologies. We expect to have four distinct eco-technologies that can be lowered down into the water. We currently envision that their designs will be inspired by, and derived from, four parent ecologies from the inshore oceans. The first ecology will include the seagrass communities that are the nurseries for so much of marine life. As I've discussed many times in this book, for years I have been using eelgrass community analogs as the basis for design of aquaculture systems that worked beautifully.

The second engineered ecology will be based upon oyster reef communities. This component of the purification system will have many different species of clams, oysters, and other mollusks. There is a technology used in marine mollusk farming known as a tidal upweller. Tidal upwellers use natural currents and tides to deflect large volumes of water through chambers housing clams and other bivalves. This increases the amount of plankton food available to the cultured organisms. We are working on adapting this technology into the *Ocean Restorer*'s design.

The third inspirational ecology is the forest-like algae community known as kelp, a seaweed common in more northern waters. We will simulate the kelp ecology with a special manufactured material that is good at attracting and holding on to diverse marine communities that filter and clean water.

The final ecological element in the complex food webs on the ocean restorer will be communities known as algae turfs. These are Lilliputian marine "forests" of attached algae made up of more than a dozen species of exquisitely shaped organisms. Algae turfs are well known for their ability to produce oxygen under daylight conditions and remove nutrients rapidly from the water.

The next step for us was to select a designer for the ship who understood both yacht design and marine biology. Laurie McGowan, the Nova Scotian described in chapter 11, more than filled the requirements and from the outset was excited about the project.

As I talked about the *Ocean Restorer*, I realized that my students were getting increasingly involved. They urged me to expand the *Ocean Restorer*'s mission to allow the vessel to double as a floating school with plenty of on-board accommodation and laboratories for students. They really liked the idea of a ship that combined restoration with ocean education. They wanted to be part of an ocean stewardship enterprise that might one day make a difference.

The drawings below illustrate the current iteration of the *Ocean Restorer*. It is a rugged little ship, with four masts and six very large sails. It's a biplane type of rig with half of the sails on each hull.

Ocean Restorer V.4.4.1, 130′ × 65′ catamaran

Underbody of the *Ocean Restorer* illustrating the framework that will house the purification systems

The whole roof of the cabin and wheelhouse area will be covered with solar cells that will generate electric power. The sails themselves will help deflect sunlight from the sky onto the cells. Windmills on the aft wheelhouse deck will provide additional electricity for its electrical propulsion systems. We plan on investigating the potential for the *Ocean Restorer* to be totally powered by wind and solar energy. Our dream is to have it operate free of fossil fuels.

I have not yet worked out a clear pathway to fund the first *Ocean Restorer* and to test at sea its living systems. The prototype will require "angel" financing of some sort and possibly the support of a strong institution as well. Nevertheless by the standards of conventional oceanographic vessels of today, the *Ocean Restorer* would be a much less expensive vessel to own and operate.

If we are successful, I can foresee the day when the seas of the world are patrolled around the clock and throughout the year with ocean restorers purifying coastal waters. From our earlier restorer experiences, we know that they will serve double duty as sea life attractors, including fish and a rich diversity of beneficial life. Like Thor Heyerdahl's balsam raft, the *Kon-Tiki,* drifting in the South Pacific to the Polynesian islands in 1947, the ocean restorer will attract marine food chains around it. For the crew and students there will be so much to see and to learn.

There is even the hope that with a fleet of ocean restorers, the Gulf of Mexico might be transformed. Such an ecological armada would be an incredible force for good. While no ocean restorer can directly treat crude oil pouring out of a well into the sea, the vessels can treat dilute levels of contamination with oil-based chemicals. A large fleet of around one hundred ocean restorers plying the Gulf of Mexico might well return the waters to their former health and mitigate the annual onslaught of agricultural and industrial chemicals that flow in from the Mississippi River. My dream is to create an ocean stewardship culture that assists the seas in restoring, for the world, their great oceanic bounty.

14

Designs for Southern Africa

In 2013 I went to South Africa for the first time with my wife, Nancy Jack Todd. Nancy was born and educated there, but I did not quite know what I would find. At the time Nelson Mandela, one of the principal architects of post-apartheid South Africa, was seriously ill. Everywhere along the roads there were signs with the words "We love you," honoring him and wishing him well. It was very moving. Somehow this man had come out of years of imprisonment and suffering with a powerful message of understanding and reconciliation between peoples that then shaped a nation on these principles. This transfer of power came without war. It was a modern miracle.

As the days and weeks passed, I observed the different peoples, blacks, whites, East Indians, and the Cape Coloured community, an ethnic group of mixed race people, working together in the technical and scientific circles which I was engaged with. I found an equanimity I had not experienced before. People seemed to share a passion for their young country.

As I was giving ecological design lectures and workshops throughout the country, we toured extensively. Most of our time was spent in the Western Cape Province, however, which included the port city of Cape Town, which is wrapped around majestic Table Mountain. We also stayed at a vineyard near the exquisite Dutch colonial town of Stellenbosch. In every direction, the horizon is ranged with rugged and jagged rocky peaks. The slopes below are mostly forested. In the valleys are neat rows of vineyards, well-ordered houses, and farm buildings. There is nowhere else quite like it.

Winery

Much of the vegetation is unique to that southwestern part of the country. Plant evolution looks as though it had run amok. One area, known as the *fynbos,* is a plant community or habitat that includes many species of the dramatic Proteaceae family composed of both beautiful flowers and flowering trees. Also dominant is another remarkable group of plants native to the area called restios, or reeds, that look a little like our bulrush family of water-loving plants.

Largely unseen, except when they surround the major cities, are the slums of South Africa. Slums represent the shadow of most societies and here they are no different. They are known in South Africa as "the informal

Fynbos plant community

settlements," a strange-sounding name for a collection of shacks crammed together without much in the way of plumbing, infrastructure, gardens, or trees.

One of the reasons we had come to South Africa was to work with several national groups including Biomimicry of South Africa; Maluti Waters, an engineering group; Mthambo Development Services; and In-formal South, a company that provided integration and leadership for all of us. As representatives of our ecological design company, we were tasked to determine if our living or ecological technologies could be adapted to the informal settlements. The projects we were to work with all involved the protection and enhancement of the important Plankenburg River, which had been degraded by pollution. There were informal settlements within the watershed that had a heavy impact on water quality in the region. We were assigned to two communities, Langrug and Mbekweni. Our task was to improve water quality within the villages as well as that of the watershed below.

Our first visit to Langrug was a sobering eye-opener. Situated on the side of a steep hill with a view of distant rugged peaks towering across the valley and overlooking a conventional community below, it was a settlement with an almost total lack of infrastructure except for electricity. A few toilet clusters and water taps were scattered through the village. Vegetation was scarce. Sewage was running down the streets and paths where small children played. The shacks were mostly made of tin sheeting, making them ovens during the day and freezing cold throughout the long winter nights. Although the rainy season was coming to an end, dampness seemed to infiltrate everywhere. The black water flowing in the eroded ditches throughout the village posed a dangerous source for many disease organisms including intestinal pathogens, which rob the health of many who live there.

I was heartened, however, by two things I observed. First was the dedication of my South African partners to help the residents of these slums. My second surprise came from some of the people I met who lived there.

Langrug in the rain

They had a commitment to making the village a better place and possessed the social and political skills to make the necessary changes happen. In this destitute-looking community, there were a few people with the savvy and skills to facilitate change. Later, at meetings in the city of Stellenbosch, I found some of the community members were active participants in a Western Cape Province government-sponsored planning meeting.

As I walked through Langrug's narrow alleys and observed the slopes eroded by the foot traffic and rains, I knew that a different kind of infrastructure was needed here. It had to substitute information for costly hardware and biological design, for machines, pumps, valves, and all the other paraphernalia of conventional water management and treatment. I began to reflect upon a landscape-based approach that would start small, at the source, and spread like a network over the landscape. It would be composed of many micro-solutions to water management and improvement. During this early stage of design, I imagined a technological landscape that would be virtually invisible to those who did not know of its functionality.

There were many reasons to think this way about design there. Everything that was not securely protected was at risk. I had heard stories of communities that had new waste treatment plants installed and of the subsequent gutting of the plants by thieves.

At first the specter of theft was extremely intimidating. In the past, we could not treat waste without air compressors and pumps. Then I realized that we would have to design without these things. We should instead look for analogs in nature and use them in our design. If we could find such analogs, they would be smarter, cheaper, more resilient, and possibly even theft-proof. They would be theoretically invisible.

The primary question was to discover what we had to work with that was ubiquitous and would underpin a new form of waste treatment and restoration of polluted water. I went back to first principles for inspiration.

First there is the Earth's atmosphere in which the air is rich in oxygen and other gases. It is all around us. We then had to find passive methods that, through subtle design, would transfer the gases into the water. Secondly, we did have abundant sunlight. A solar solution would allow plants to act upon the wastewater through photosynthesis. There are some higher plant species that can pump oxygen directly into the water through their roots. Also, most plants manufacture saps in their roots. These in turn feed the microbial life that acts beneficially with the wastewater and transforms it. And the hillside itself could function as an energy source. We could let gravity rule and have the slope move the water. Our design challenge was to tap into these energies and utilize them to help heal the water.

At this point I turned for help to the study of flowing systems as they appear throughout nature. This led me to thermodynamics, the science of how to convert heat or other forms of energy into work or, conversely, work into heat. The first law of thermodynamics dictates the conservation of energy, and states that energy can be transformed from one form to another but cannot be created or destroyed. The second law states that the entropy of an isolated system never decreases because isolated systems always evolve toward thermodynamic equilibrium, which depends on maximum entropy. This law predicts, for example, that heat flows from a region of higher temperature towards a region of lower temperature and not in the reverse direction. The same is true of any energy source. This is referred to as the one-way flow principle.

I was looking for patterns in nature that could shape the most efficient interactions between water and air or other substrates, ways in which life itself interacts with other forms of life and the environment. I was curious why the similarities between flow patterns of the deltas of large river basins are not unlike the shapes of trees or the patterns of organs and flow systems within our own bodies. There seems to be a unity in the patterns found throughout nature.

In my research, I discovered the work of Adrian Bejan. He has articulated a new law of physics that defines the pattern that connects the inanimate realm to the living and beyond to encompass the known universe. In his book, *Design in Nature* with J. Peder Zane, he declares there is a universal design law that governs and shapes flows of all kinds. There is a connection between diverse patterns such as the design of a lightning bolt, the shapes of watersheds, the human lung, arterial systems, the brain, and the branching patterns in trees, and, in another direction a connection to human transportation and information networks and the shaping of cities. It is about flow and flow patterns. Conceived in 1995, he called his law "the constructal law" or the law of design in nature. His seminal book, published in 2013, is an exploration of the law and its influence on the world around us and the world we are shaping each day.

Bejan is changing the rules for how we think about design. First, he declares that what our minds perceive as design, including configurations, rhythms, and scaling rules, is present in all flow systems in nature. His seminal idea is that up until now design phenomena were not covered by the existing laws of physics. The first law of thermodynamics commands the conservation of energy, the second that flow should move from high to low. His is a third law, the constructal law, which commands that energy movement should flow in configurations that become easier and more efficient over time. What he is saying is that if physics is to cover nature completely, it must be endowed with an additional principle that accounts for the phenomenon of design generation and evolution everywhere and in everything. His third law accounts for this.

The constructal law predicts that evolution occurs because the tendency of all flow systems is to generate better and better designs for the currents that flow through them. In his view this improvement applies not only to river basins and forests but also to biological creatures from single-celled organisms to vertebrates, including humans. There is a tendency in nature to generate shape and structure to facilitate flow access.

The constructal law also predicts that our circulatory system should have a treelike structure of round tubes with a few main channels, our arteries and veins, and numerous tributaries, our capillaries to deliver water, oxygen, and useful energy to every cell. Bejan claims that the evolutionary history biologists have charted and the series of adaptations they have detailed are all expressions of the constructal law. Further, these same tendencies and dynamic processes apply to inanimate systems as well.

This discussion of the laws of nature has a bearing on the design of technologies for waste treatment and, most importantly, ecological technologies for water improvement. Water quality improvement requires bringing polluted water into intimate contact with a diversity of living organisms and the complex biochemical interactions that transform polluted into pure water.

Could we design systems for wastewater treatment in South Africa that would include the genius of nature like dendritic forms or shapes that seem to be almost universal, whether in the structure of a tree, a bolt of lightning, or a nervous system? They all reflect movement and the optimization of flow and contact with surfaces. Might we then engineer technologies that could transform the quality of water from an unwanted state like sewage to a preferred one like pure water? Might dendritic shapes and design become the basis for the invisible technologies we planned for South Africa?

As I mentioned before, in the informal settlement of Langrug, wastewater flows through the paths and streets in rivulets and along shallow concrete channels. There are also partially buried buckets along the paths into which individuals pour wastewater, especially gray water, after use.

I created a dendrite design that would transform the whole village into a treatment complex beginning with micro-catchment elements throughout the community. Our ecological engineer designed the whole system from the receiving buckets on the front end to new flow patterns in the village. They are combined with living technologies appropriate to each stage along the way. His design ended with an eco-machine adapted both to the site and to the future economic needs of the community.

The drawing illustrates the overall scheme from the top of the community on the hill to the bottom, which is adjacent to a sports field. The pattern is one of narrow branches that come together, first at small nodes that then undergo further convergence at larger nodes and so on throughout the community. They represent pathways for the wastewater through small diameter piping down the slope. The capillary-like design allows for new connections to be added in the future if the village grows.

The small circles represent waste collection points where wash water is carried by the residents to buckets partially buried in the ground. These collection and filtration systems are central to a flexible design and the early stage of waste treatment begins within them.

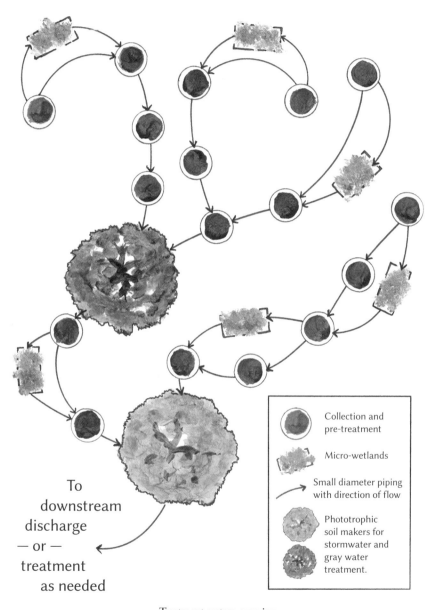

Treatment system overview

Collection system

Our plan was to prototype different kinds of collection buckets. Some will have modified slow sand filters over wood chips and compost for active treatment. At the other extreme we will compare their performance and practicality with filters of shade cloth for screening.

On the schematic, the rectangular boxes represent a side stream technology. They are miniature constructed wetlands designed to expose the wastewater to contact with the roots of water-loving plants including cattails and bulrushes as well as native reeds. The drawing shows that the micro-wetland systems are on a side stream so that if the flows become too great during rain, most of the wastewater will bypass the micro-wetlands and stay in the main channel.

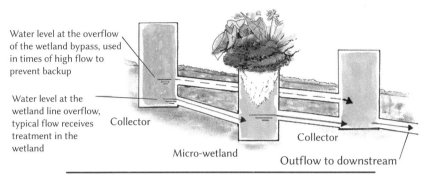

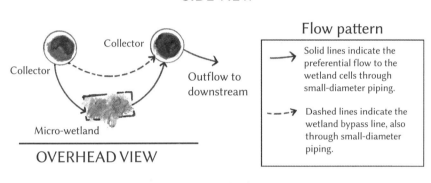

Micro-wetlands

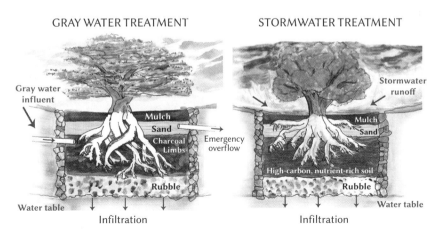

Tree wells: phototrophic soil-maker technologies

The large central nodes are a new technology. We have called them "phototrophic soil makers" because they include trees and convert the organic matter in waste to carbon-rich soils. They also combine the culture of fruits in orchards with rapid soil formation and waste treatment. They should help stabilize the community ecologically in a diversity of ways.

Basically, the technology starts with the digging of a deep pit or well into the ground. The pit will contain a variety of materials that will allow the waste to be treated and after treatment infiltrate into the water table below. On the bottom will be a layer of rubble with void spaces that aid the movement of water. Wastewater flows into the pit below the surface. There is an emergency overflow for rainy periods when infiltration into the ground is not rapid enough. The overflow will flow downstream to the next central node and the treatment will be repeated.

Above the rubble, the well is filled with the "soil-making" media and will include a blend of charcoal, seashells, wood chips, and compost inoculated with

Tree well technology being installed

Prototype tree well treatment system

a diversity of fungus species and soil organisms including earthworms. The high carbon base of the media will ensure the entrapment of nutrients from the wastewater and support the soil formation processes. Into the media we have planted orchard and nut-bearing trees adapted to the South African climate.

The trees are essential to the technology. They will assist in keeping the soil porous and aerated and their root saps will catalyze the formation of soil within. The top layer in the well will include a mixture of sand, mulch, and compost that in turn will control any negative odors from the well.

The wastewater treatment tree technology will go a long way towards reducing flooding and erosion in the community. It will also modify the climate, cooling in the summer and warming in the colder months. Furthermore, when bearing fruit, the trees will add greatly to the diets of the local people.

It is expected that sewage water will increase as more toilet and washing pavilions are developed in the future. To serve this need, we have proposed an eco-machine to treat sewage, gray water, and rain runoff as well as excess water that might get through the wetland and tree technologies on the slope.

The design was a radical departure for us. It is a formal example of a dendrite approach to design. Max Rome, the engineer, created a tightly packed, cellular design with cascading flows that allows the water to flow in several directions, but always downhill. In the preliminary design, there are six levels or elevations composed of twenty-six cells. Between the cells are gutter-like structures that spread out the flowing water into thin films. The water, when exposed to the atmosphere as it cascades downwards, gains a modest but significant amount of oxygen from the atmosphere. With the large number of cells and the variety of flow patterns between the cells, the system's biological diversity should be great. The physical structure, combined with a richness of species within, should allow for a wide range of biochemical pathways available to treat the wastewater. The hydraulic retention time, a measure of the time the water remains in the system, will approach seven days at average flow rates. This should provide sufficient time to significantly reduce pathogen levels in the water.

This design, with its absence of pumps and mechanical aeration, should come close to the "invisible" technology we had hoped for at the outset. It will look like a floating garden at the base of the hill. It will be beautiful.

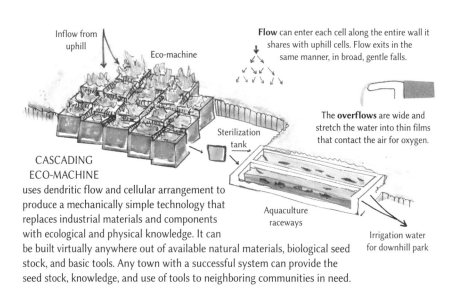

The South African eco-machine

Since water is scarce for much of the year, we plan on developing ways for the treated water to be reused for economic purposes. One option is to grow attached algae species on screens placed in shallow raceways and harvest the algae to feed herbivorous fish such as tilapia and other species that are native to Africa. Other options being considered include a system to hydroponically cultivate valuable trees and shrubs within thin films of flowing water. The growing of valuable cut flowers is another possibility being considered.

The illustration on the left depicts the community as it is today. It is barren and devoid of much vegetation. The middle illustration shows the community five years after the piping, waste separation, and the installation

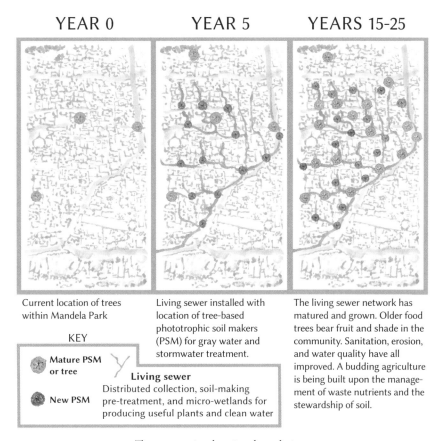

The community changing through time

of micro-wetlands and the tree-based waste treatment technologies have been initiated. In five years, the impact of the vegetation on the community would be already apparent. On the right is the community fifteen to twenty-five years from now. The area is more park-like and bountiful. The photos and sketches show the considerable progress that has been made.

What is important to note is that the whole design process embraced the notion of local economic development as part of the overall water management plan. What will fit best into the economic life of the community has not yet been determined. Ecological design requires several notions from ecology itself. The first is the ability to adapt to a changing environment and circumstances. The second is the idea of change and evolution. This is what ecologists call succession. The landscape is transformed through time. It becomes more diverse, more inclusive of people and more economically robust. In the wild, given sufficient moisture, it takes approximately a quarter century for a bare field to become a young forest. It will take a similar amount of time to transform a hillside community built on rubble to its full potential as a garden village that supports the people within.

15

Appalachian Spring

Appalachia, as most Americans know, is a broad cultural area in the eastern United States that ranges from the southern tier of New York state to northern Alabama, Mississippi, and Georgia. It is one of the most biodiverse temperate regions in the world. It is home to 6,300 plant species and is a global hot spot for freshwater aquatic life. In Tennessee there are 290 species of fishes, more than in all of Europe. The method of mining used there, involving huge machines, has literally destroyed over 1 million acres of one of the most biologically diverse temperate forests on Earth.

A decade ago before I started writing this book, the Lewis Foundation, committed to a viable future for Appalachia, asked me to study the environmental impact of mountaintop removal and valley-fill coal mining throughout the region. With a small team of my former students, I set out to review the scientific literature. I read all the books I could find on the subject and viewed aerial photographs showing the unbelievable destruction that has been the fate of this economically impoverished region. Our task was to determine the severity of the impact in biological terms of the practice of mountaintop removal followed by the burying of valleys and streams.

I was not prepared for what we found. The horrors of such mining practices boggled my imagination. We saw endless scarred rocky surfaces where there had once been vibrant forests and healthy soils. We learned of the destruction and burying of streams, the loss of native species, the breakup of mountain communities, and of huge earthen dams filled with toxic coal slurry. We saw photographic images of a school evacuated after being buried by waste from a failed dam. The list of horrors went on and on. My colleagues and I began to doubt the usefulness of our work. Would it ever be relevant? The coal industry had dominated the region since the early twentieth century. Alternative futures were in short supply. How could our studies influence the course of events?

I asked the foundation if they would let me change course. Would they allow my ecological knowledge and perspective to search for an alternative scenario to Appalachia's future? They said yes to my suggestion. I knew the path forward would not be easy. It would involve interlocking issues of ecology, agriculture, forestry, economics, ownership, education, and institutional underpinnings of culture and politics. The work would be on a scale I had never attempted before.

To focus my mind and give the future a literal image, at the outset I sat and wrote down the story directly below where I attempted to envision the future. Using the narrative, I was then able to outline the design parameters and the path ahead. Imagining the future was exciting and opened up a much broader scope for the project. Once written, I knew in broad terms the path for my enquiry. The story became a crucible for design.

A VISION FOR THE FUTURE

Imagine a time in the future. You are flying silently at a low altitude over an Appalachian valley in an electrically powered ultralight plane whose energy and propulsion are derived from solar cells incorporated into the fabric of the wings. Everything below you is green, but in the greenness there is both pattern and variety. In recent years, patterns and variety in the vegetation have been created anew as the result of human-orchestrated biological restoration on the ground. The last time you saw this land was several decades ago when it was scarred and laid barren by mountaintop removal and valley-fill surface methods of coal mining. Then it looked like a moonscape, devoid of life and people. Today it is different. You notice that the landscape below has both block and contour patterns. Some of the blocks and contours are composed of trees. In some, the trees all look the same and in others the tree types are diverse and have different shapes. Trees, orchards, vineyards, and nut groves dominate the slopes.

Deeper into the valley other blocks are fields made up of waving grasses, grains, and pastures that reflect in their colors the diversity of the farms below. In some of the pastures there are livestock. Cattle, bison, and goats graze in well-fenced paddocks. Fat active pigs are seen scurrying in and out of oak groves in their search for abundant foods.

The other big change is the number of people in the landscape. There are lots of them. Once there were a few houses, but they were eventually washed away by floods caused by rain, bare slopes, and the collapse of a coal slurry dam. They have been replaced by a thriving town, much larger than was typical when you were last there in the late twentieth century. A closer look shows other changes. The sheer diversity of activities in the town is startling. There are libraries, a hospital, churches, galleries, shops, civic centers, and the pride of the town, new schools in the heart of the village. Integrated into its fabric are clusters of buildings in which manufacturing and a variety of services take place. They supply not only the area, but in some cases their products are sold throughout the world. The factories don't pollute the atmosphere or the water, so they have been situated where people live and are easy to reach. In the heart of the town is a park defined by a stream that runs through it. The park honors the stream and the surrounding landscape that helps keep the waters pure. In its center, there is a memorial to honor the people who had the courage and vision to transform the region into an ecological and economic jewel. The list is long.

As you bank the plane to have another look, you see that the town is connected, up and down the valley, with three parallel ribbons of transportation that are for cars, light rail trains, and walking and bicycling paths. Your plane climbs to look for the airport. On the way, it passes over a cluster of slowly turning windmills on a hilltop ridge and then descends into the valley. You have come to visit the people who made your plane and to observe firsthand the efforts of their research and development. They have developed ultra-high-strength composite materials from local woods that are grown by forest farmers and they are making the area famous for high-quality trees. They are already using the new composite materials in the latest carbon-neutral tractors and pickup trucks. It's a very busy place actively developing a plethora of new ideas. You land the plane on a grassy airfield and set out to meet the workers. You discover that the firm is owned by the people who work in it, as are many of the other companies in town. You also learn that the town, as well as the county in which is it situated, has invested in most of the businesses. The town is an ownership community.

You walk downtown to the inn by the river. There is a bridge next door where many of the children are casting for trout. Within the town limits only children are permitted to fish in the streams. The meal that evening, even including the wines and the walnut cooking oils, all came from ingredients produced in the valley. The town is now famous for its wild-caught and locally grown and processed foods.

At dawn, you are ready to depart. Like a soaring hawk, you lift off with the sun and fly toward the west. The region has been transformed and you have seen the future.

• • •

For months, I worked on a successional model for healing and rehabilitation of Appalachia. I was guided by a theoretical framework that uses dynamic change and maturation of living systems in the wild, formally known as succession. Nature, when given the opportunity, rapidly transforms itself from a barren setting into an increasingly complex system that over time supports increasing biodiversity and abundance. A typical example is a bare field that has been bulldozed. If left alone, and if there are occasional rains, the area is quickly seeded by pioneering weeds, which set the stage for perennial plants. As the years go by, these plants are replaced with shrubs, then trees. Eventually a forest returns, bringing a bounty of wildlife with a diversity of life forms that are coupled by symbiotic mutual relationships. Water cycles are reestablished, streams return, springs rush out of the ground, and the local climate is stabilized. Where there is healthy water, life abounds.

Succession became the template for my design process. Within this successional framework, I posited the notions of the First Order, Second Order, and Third Order Ecological Design models outlined in chapter 3. First Order Ecological Design represents techniques and technologies applied to the landscape. Second Order Design is the linking together of processes and practices into new associations and entities. Third Order Ecological Design addresses institutional structures and their changing relationships over time. Institutions have their own succession, and each stage in a healing landscape has its own appropriate mix of entities serving both individuals and the common good. Third Order Design transcends the usual left-right political frames of reference and addresses the needs of very complex systems and their evolution.

The product of my design enquiry into the future needs of Appalachia was a report to the Lewis Foundation entitled "A New Shared Economy for Appalachia: An Economy Built Upon Environmental Restoration, Carbon Sequestration, Renewable Energy, and Ecological Design." The report was condensed and then submitted to a newly announced competition, the Buckminster Fuller Challenge. It was to be an award for the "best idea to help save humanity." My submission won the award.

My approach to restoring the coal-mined land in Appalachia was predicated on a simple, fundamental idea, namely that soils define societies and that the quality of their soils dictate to a high degree their durability and fate. My effort would revolve around creating deep, rich soils even where there was rubble and rock. The plan was to explore ways to create fertile soils where there were none.

The strategy was soil creation over vast acreages. The emerging soils would sequester carbon, help stabilize climate, and underpin landscape restoration, intensive agriculture, animal husbandry, and agroforestry on lands where before mining had supplanted forests. My colleagues and I would have to jump-start succession on the landscape by selecting hardy plant species that have enhanced soil-building qualities. In the beginning these plants would need support with techniques I had combed the world's literature to find. Fortunately, I had worked with many of them in the past and had confidence that in combination they could support the rapid formation of soils.

The first technique in our investigation was remineralization through the addition of finely ground-up rock powder, which is sometimes called rock flour. Finely ground powders from different parent rock types can provide trace minerals and substrates for beneficial bacteria and fungi on even the most barren surface. While working on the Appalachian report, I carried out an experiment on badly overgrazed, burned, and eroded slopes situated in a seasonally arid area of Costa Rica. My associates and I had carried out a reforestation experiment involving replanting primarily with native tree species. Half of the trees were planted in holes containing aged manure and half were planted in holes with the aged manure that was combined with rock powders. Such was the difference that the seedlings with rock powders did much better than the manure-only trees, so much so that the caretaker of the trees decided to use rock powders with every tree. She felt healthy trees were more important than experimental comparisons. It was hard not to agree with her despite the consequences for our trials.

The second method that seemed pivotal to us was to spray beneficial organisms onto a newly planted area. I have found, as have many others, that brews of earthworm compost and regular compost, aerated and cultured in open vats, can produce communities of beneficial bacteria and fungi. When they are sprayed over soils and crops, the response of plants can be dramatic. They grow more rapidly and fend off diseases more readily. This technique can be scaled up to treat large areas cost effectively. I have used several companies making commercial microbial sprays, including those from Urth Agriculture.[1]

When excess organic matter is available, a third soil-building method is composting. Although treating large areas can be quite difficult and expensive to manage, compost can be very important. Compost islands spread throughout a landscape can act as reservoirs of organic matter and can enhance soil formation locally.

A fourth technique for aiding soil formation involves the addition of a type of charcoal known as biochar onto the land. Biochar is produced by burning tree trimmings and other woody matter at relatively low temperatures under reduced oxygen conditions. The result is a dark material that can function as a nutrient reservoir and a stable carbon source in the formation of dark earth soils. I personally have no direct experience using biochar, but in Costa Rica I have observed its effectiveness in agriculture first-hand.

Finally, there is the possibility of using chemical fertilizers to jump-start support for newly planted crops. When we're planting on bedrock or compacted soils, a one-time use of fertilizers may support the initial growth of plants that help trigger the formation of new soils.

The real test lies in whether we can restore landscapes from scratch with a relatively modest investment. I knew from our research and experience that forests can be created on lands that have been mined, but the costs can be quite prohibitive.

Mined landscape

Appalachian foresters from Virginia Tech's Powell River Project have been studying ways to recreate new forests on old mining land for many years. Their work is groundbreaking, both literally and figuratively. They have created a gold mine of reforestation information, especially around the use of suitable species. However, what they found was sobering to me. The depth of the soil was critical. Growing media twelve inches deep had a hard time supporting newly planted trees. When trees were planted into a four-feet-thick medium that included dirt added to the mixture, trees did well and their economic value was twenty-eight times greater. With the thicker media or "soil," the foresters found that the return on their investment was around 10 percent based on the value of the trees alone. This need for thick growing substrate was a stumbling block for our enquiry. Earth-moving costs would stymie any large-scale efforts to reforest the region.

To their great credit, the Powell River researchers developed a successional or ecological model for establishing soils and woods. In my view this was an important legacy in the annals of the region. They developed a groundcover plan that was initiated by planting annuals that dominated the soil for the first few years. This was followed by planting herbaceous legumes and deep-rooted perennials that dominated for the next three or four years, to be followed by hardy pioneering nurse trees or biomass trees that would dominate toward the end of the first decade of plantings. After fifteen years, the landscape could be transformed into a young forest featuring valuable trees including timber trees such as pines, oaks, ash, and maples as well as orchard trees producing crops each year.

Our objective was to find some way that organisms, especially higher plants and their symbionts or allies, could be used to start the soil creation process. There was no way that we could quickly or cost-effectively create deep growing media to plant trees into the ground. Our approach involved seeking out plant species that had aggressive root systems and were tough enough to penetrate compacted rock substrates and to mine what they needed from the soilless environment. It seemed our hope might lie with warm season perennial grasses. The species we selected can dominate in open savannah-like landscapes, and included switch grass *(Panicum virgatum)*, big bluestem *(Andropogon gerardi)*, and Atlantic coastal panic grass *(Panicum amarum)*.

In the past, mine site reclamation has involved the seeding of non-native and invasive species of annuals, sometimes in combination with legumes and fertilizer applications. The result is usually an eroded landscape with sparse vegetation and extremely shallow soils.

A doctoral student of mine at the University of Vermont, Samir Doshi, did a warm season grasses experiment in a plot at the Powell River Project. On a five-acre test site he grew the three species of warm season grasses. Half the test plots were fertilized with nitrogen, potassium, and phosphorus. The plots were inoculated with mycorrhizal fungi that developed a symbiotic relationship with the roots of newly planted grasses.

Overall the results were very encouraging. One species, the big bluestem, did not grow on any of the fertilized or non-fertilized sites. However, the Atlantic coastal panic grass and switch grass did well the first year. At the end of that year the fertilized Atlantic coastal panic grass yielded 302 lbs/acre (336 kg/hectare) and the switch grass 213 lbs/acre (237 kg/hectare). The unfertilized Atlantic coastal panic grass yielded 20 lbs/acre (229 kg/hectare) and the switch grass 140 lbs/acre (156 kg/hectare). Two years later the grass yields, both fertilized and unfertilized, increased by over an order of magnitude. Atlantic coastal panic grass that was fertilized yielded 4,005 lbs/acre (4,450 kg/hectare) and 2,907 lbs/acre (3,230 kg/hectare) without fertilization. The fertilized switch grass yielded 4,019 lbs/acre (4,465 kg/hectare) and the unfertilized plot 1,740 lbs/acre (1,933 kg/hectare). In this brief two-year period the two grasses, with or without fertilization, in partnership with their arbuscular mycorrhiza fungi had begun to make soils in this most inhospitable terrain. These fungi are closely associated with the roots of the plants. The mycorrhiza fungi work by penetrating the roots of plants and providing them with nutrients and water in exchange for photosynthetically derived sugars and other useful chemicals from the plants themselves. The relationship is one of the essential biological pathways for the formation of soils.

Over the two years of the experiment the switch grass and the Atlantic coastal panic grass increased the organic matter in the soil significantly. The highest increase was 46 percent in the unfertilized Atlantic coastal panic grass plots. While this information needs further corroboration, it appears that some warm season perennial grasses can adapt to damaged landscapes and trigger the formation of soils. This fact became the basis for our restoration plan for the over 1 million acres of coal-mined lands in Appalachia.

Perennial warm season grasses growing on the test site

The hardy perrenial grasses became a lynchpin in our thinking. Besides being critical to the early formation of soils on the landscape, they have many other positive attributes that can be applied to carbon-neutral and viable economic solutions for the region. They are suitable for forage production and cover for game, as well as being useful for the phytoremediation of pollutants in toxic sites. After harvesting they can be used to produce fibers, electricity, and heat through combustion. If managed wisely, they can function to sequester atmospheric carbon dioxide and help stabilize local climate. In recent years it has been found that these grasses can be used as biomass crops for the fuels ethanol and butanol. Perrenial grasses can be converted to ethanol with an overall efficiency of around 35 percent. As a consequence, the grasses became the heart of our plan to transform the region.

At the very beginning of the restoration process, our plan would involve planting an area big enough to support a local economic component almost from the outset. My hunch is that a thousand acres might be a suitable size at the start. The area would be planted entirely to warm season grasses and innoculated with soil organisms including bacteria and fungi. Other imported materials could include mineral enrichment with finely ground rock powders.

The grasses would be harvested and the crops would be used to produce biofuels as an early cycle revenue generator as well as for building soils and sustaining livestock on the landscape. Over the ensuing years as soils form, the diversity of biological activity would increase to include other crops and to trigger ecological and economic succession on the land.

Towards the end of the first phase of the project, our plan includes the widespread planting of willows and poplars for biomass production among the grasses. Their biomass could be converted to the production of biochar and be used for the local co-generation of electricity. Many of the willows can be harvested through coppicing and therefore not killed. Every four to six years the trees are cropped. Afterwards they would resprout and grow again.

By the eighth year, the grasses over much of the area would have been interplanted with the beginnings of a forestry landscape that would include fruit, fodder, and nut trees as well as vine crops with wine grapes. Grasses would still produce biofuels. Biochar would be a byproduct of the cultivation

Appalachian Spring

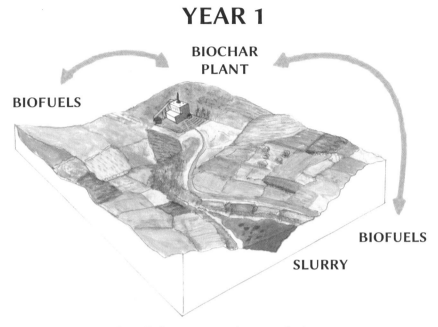

Year 1: Early succession on the reviving landscape

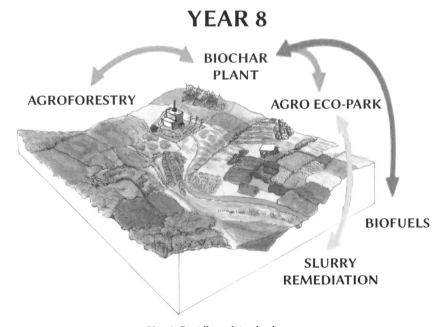

Year 8: Rapidly evolving landscape

of biomass willows and hybrid poplars. The land will be beginning to be fertile enough to underwrite an agro-eco-park that will convert the primary productivity on the landscape into value-added crops. By this stage the land is able to support the restoration of adjacent lands with seeds, microbial amendments, and soil-enhancing biochar. All the while, atmospheric carbon is being stored in humus-rich soils as well as in standing vegetation.

By the sixteenth year, the original site would support a wide diversity of activities. The toxic coal slurry, a byproduct of coal mining, would have been cleaned up using a combination of eco-machines and constructed wetlands linked together hydraulically. Water would be circulated between them using wind or solar energy.

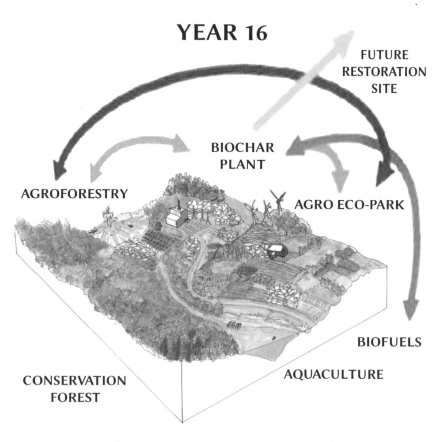

Year 16: Mature succession and an expansion to new terrain

By the sixteenth year, when the job is done, the environmental cleanup technologies can be converted to commercial fish farming. When natural springs have returned to the landscape, as they undoubtably will, they can be employed to support trout culture. Ponds, designed to capture and hold rainwater, could be converted to support the cultivation of catfish and buffalo fish. By this time, the biomass tree plantations will be at peak production. The orchards and vineyards in the sixteenth year will be producing bountiful harvests and will employ an increasing number of local people. Valuable nut trees, such as walnuts, will be managed for timber as well as their nuts. Livestock including cattle, sheep, and pigs will be part of the landscape and will graze in rotation amongst the tree-dappled pastures.

Perhaps the hallmark of the enterprise in its sixteenth year will be its ability to underwrite new restoration enterprises. Training and skills developed by local people on the early site will provide the talent pool for the rapid expansion over tens of thousands and eventually hundreds of thousands of acres of former coal lands. What will be needed at this stage is a successional model for the whole economy, a working example of Third Order Ecological Design.

I have spent many years trying to figure out a model for the large-scale restoration of the environment. There is no simple solution. There are issues of land ownership, financing, education, management, and governance. My goal in all of this includes the dream of giving the working landscape back to the people of the region. At some level, lands and natural resources should become a true commons in which all people share in the bounty in one way or another. This leads me to the idea of a cultural succession for Appalachia that is the equivalent in its own way of the ecological succession going on within the landscape itself. Not being politically or culturally trained, I sat down and drew an image of what the process of creating a viable commons might look like. The only way I knew how to proceed was to visualize the landscape and its changes through time for the people that live on the land or derive their livelihood from its resources.

Taken together it may look like a political system, but as an ecologist, I see it as a symphony of relationships that must change through time. Our whole landscape is evolving and becoming more diverse as well as more useful to society as a whole. Below is my concept. It's not a blueprint or a roadmap but an idea that can guide us in creating new economic and social relationships out of our efforts to restore the lands and inform the Earth's stewards.

SEVEN SUCCESSIONAL STAGES TO REGENERATION

I have illustrated seven successional stages and have given them each a life span of twenty-five years, roughly equivalent to a human generation. Each stage might be longer in practice, maybe even fifty years; we simply do not know the precise amount of time involved in the whole process because the driver of time is the evolving landscape itself. The beauty of this is that

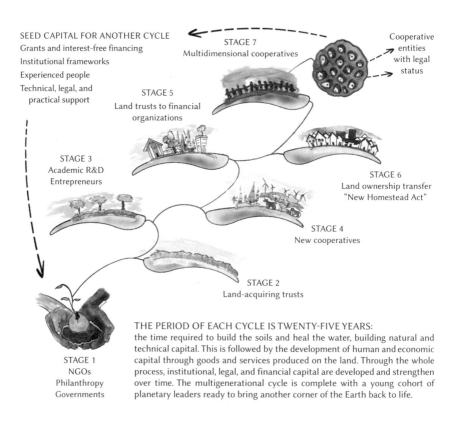

THE PERIOD OF EACH CYCLE IS TWENTY-FIVE YEARS:
the time required to build the soils and heal the water, building natural and technical capital. This is followed by the development of human and economic capital through goods and services produced on the land. Through the whole process, institutional, legal, and financial capital are developed and strengthen over time. The multigenerational cycle is complete with a young cohort of planetary leaders ready to bring another corner of the Earth back to life.

Institutional succession on the evolving landscape

it becomes an economic model very early on; each stage adds complexity, including additional economic elements.

Stage 1 is about people coming together to plan the undertaking of large-scale landscape restoration. These are the change agents and visionaries. They will come from many backgrounds. What they have in common is the desire to transform the region. They may come from, or be connected to, nonprofits or NGOs, or they may work with philanthropic entities. Many will represent local, regional, and even national organizations. Young people will be involved. Taken together, they have the political will to directly shape their world to their collective vision.

Stage 2 could involve the creation of land trusts that use the connections formed in Stage 1 to raise the money to acquire former mining lands. It is possible that progressive mining companies will donate land to the enterprise. The land trusts will need the financial resources to begin planting hardy perennial warm season grasses to trigger soil formation. One important source of long-term funding may turn out to be the newly developing carbon markets that are a key element in global initiatives to control climate in the future. These days it is possible to measure the amount of carbon being stored in soils and to determine its value precisely. The more effective these projects become at sequestering and storing carbon in soils and biomass, the more valuable they become to the world at large.

Stage 3 is the research and design stage in which universities and academics study the landscape to optimize the selection of ecological and economic strategies for the land. This in turn should launch an entrepreneurial phase with startup companies that would establish an infrastructure for the production of biochar and liquid fuels as well as the detoxification of mining residues including coal slurries. Finally, the new agrarians, who are the backbone of the restoration effort, will need to establish themselves into meaningful associations. This phase also will require bringing seed capital into the process.

Stage 4 will see the creation of new Appalachian natural resource corporation or corporations. Many will be employee-owned and locally managed. The companies will be empowered to carry out activities on the land that will no doubt be shared with other organizations active on the same land. This approach works for aquaculture in the inshore marine environment. Coastal states make available grants to individuals or companies for the exclusive use of a section of the inshore ocean. Oyster and other shellfish farmers as well as seaweed growers work this way.

An alternative approach may be selected in which new corporations operate in a place-specific context and undertake all the various functions on the land including producing valuable byproducts. In this model, it will be determined what is the minimum acreage necessary for a multifunctioning corporation. At the other end of the scale it should be ascertained what the optimal size of a land-holding entity might be.

Stage 5 is the stage in which the land trusts increasingly become financial institutions. The land is now producing revenue that is flowing back to the trusts. They can, and probably should, become mortgage or lending institutions. Their role now will be to provide financing for local ownership of most, or all, the natural resource assets.

Stage 6 I am calling a "New Homestead Act for Appalachia." Those eligible for support will have already paid their dues working on the land in the early years. They will have experience and newly acquired land skills. In a sense, what I am proposing will give the people of the region access to financial resources within a landscape restoration setting. The homestead act will be tailored to the land-based models that are chosen. If the model is to treat the land as a commons so that it is used by diverse and hopefully symbiotic economic activities, then the homestead act would give people of the area stock or equity in the corporations in charge of the stewardship of the land. Or if the model that is chosen includes private ownership under a scheme where certain ecological rules, such as carbon capture, must be followed, then the financial instruments would look more like mortgages.

Stage 7 involves the creation of cooperatives. Their primary function will be the life-long education and skills accumulation of the people of the region. Cooperatives are currently playing an important role in the emergence of natural resource–based enterprises in a number of states in the United States. Take, for example, a willow growers' cooperative in New York, a switch grass cooperative in Iowa, and an alfalfa producers' cooperative in Minnesota. New cooperatives are currently providing viable business structures for a variety of biomass-related enterprises.

The legal basis for cooperatives was created in 1926 by the Capper-Volstrand Act. The law stated that the cooperative owners must also be users of its services and that the cooperative must be democratically controlled. Each member has one vote and the benefits are distributed to the extent to which a given member uses the cooperative. A member that delivers twenty tons of biomass annually, for example, might have twenty shares, but still only one vote. One share of common stock is given to each member to assure a voting privilege. Preferred stock can be purchased, but does not result in additional voting privileges. Federal law limits returns on preferred stock to 8 percent annually. Cooperatives can secure capital through a variety of mechanisms, including the Farm Credit System. Currently the Farm Credit System provides about $304 billion worth of loans, leases, and related services to rural enterprises throughout the country.[2]

There are three types of cooperatives: marketing, purchasing, and service-providing cooperatives. Many cooperatives provide all three services by marketing products, purchasing the inputs needed by the producers, and providing such services as insurance, credit, and technical management as well as skills training to its members. What has not, to my knowledge, been explored by the cooperative movement is the idea that they provide the social glue of a community by being centers of education encompassing all aspects of carbon capture and natural resource management. I can imagine them being connected to local schools from the early grades on through colleges and universities. If the restoration project involves life-long education, then there is a very good chance that it will help form a strong, resilient, adaptive, and long-lived economy. This economy has the potential to support the land, provide meaningful work, and strengthen the arts and the culture of the region.

My Appalachian journey began with my studying the horrors created by large-scale surface coal mining and the smothering of valleys in the region. Over a million acres of incredible and bountiful forests have been destroyed and many of the people of the region have been forced to leave or continue on living impoverished lives. The tragedy is beyond measure. What I learned along the way is that this need not be so. Nature, life itself, has powerful recuperative powers. Even in the most ravaged places, there is reason for hope. We need to ally ourselves with the diversity of life and each other. Healing the planet is then possible.

16

Re-Greening the Earth: The Challenge of the Sinai Desert

As our concerns deepen for the fate of the Earth and its people, we urgently need large-scale initiatives to reclaim ecosystems. This book has been an exploration of my efforts in that arena, and I'd like to close with some final examples of pioneering techniques and technologies to reverse environmental degradation, as well as a closer look at work to restore parts of the Sinai Desert with a group called the Weather Makers, which shows that even desert landscapes can be re-greened.

The reclamation movement was started by conservationists, inspired in part by Aldo Leopold's remarkable book *A Sand County Almanac,* first published in 1949. Leopold was a forester and an ecologist with a deep passion for nature. He understood that an ecologically healthy landscape has integrity, meaning, and value. Leopold's writings struck a deep chord with millions of readers.

Another of my favorite characters in the restoration movement is Alan Savory, a wildlife biologist and farmer from Africa. Raised in Southern Rhodesia, now Zimbabwe, his insights were derived in part from his powers of observation of nature. His story is told in his book, *Holistic Management.* Like myself, he is a winner of the Buckminster Fuller Design Challenge. Alan Savory had insights into the great partnerships between the plants of the African grassland, the grazing animals, and the predators who preyed upon them. He discovered that the robustness of these great landscapes needed both the grazers and their predators in dynamic balance for the health of the plants and their soils. If the predators were killed, or the diversity of grazers reduced, the vegetation would suffer. The plants needed the grazers to keep moving across the landscape as if they were being pursued by predators. These associations were dynamic and complicated. When the relationships were tampered with, the land began to die. To his great credit, Savory used his observations to create livestock farming systems that mimicked wild ones. He learned that when done correctly, with cattle moving from place to place as if they were being chased by predators, the presence of domestic animals could also restore degraded lands. His ideas have been employed widely throughout the world. He has taught us the importance of biodiversity; when living systems are healthy all the different kingdoms of life are true partners.

Another important and expanding agricultural conservation and restoration movement is known as *permaculture*, meaning permanent agriculture. Permaculture integrates food crops, forestry, animal husbandry, and wild plants into a close holistic association with the landscape. Originally developed in Australia in the 1970s, its founders were biologist Bill Mollison and his graduate student, David Holmgren. Their first book was *Permaculture One*. It was followed by Mollison's *Permaculture: A Designer's Manual* in 1988. The latter book is a veritable feast of ideas for practitioners. It covers agricultural landscape restoration and successional ideas from ecology when applied to agricultural landscapes. The permaculture movement has since matured and spread around the world. It is widely taught and the current generation of practitioners is expanding their reach and influence. Ben Falk, a writer based in Vermont, offers a wonderful exploration of the agricultural meaning of place in a northern climate in his book *The Resilient Farm and Homestead*.

Janine Benyus and her book *Biomimicry: Innovation Inspired by Nature*, first published in 1997, have had a large influence on the design field, especially industrial design. The book's basic premise is that there is genius in nature that, once discovered, can shape all kinds of new technologies in many fields from ceramics to aircraft design. It differs from the work of Savory and Mollison described above in that it focuses on how individual species solve their survival problems efficiently and elegantly. It applies this knowledge to the design of technologies and more recently to landscapes.

More recently *biomimicry* is becoming an environmental restoration science and practice. Biomimicry is creating an ecological extension of itself that can be applied to agriculture, carbon sequestration, and habitat restoration. Biomimicry concepts are appearing in whole systems design. The emerging field has been called eco-mimetics. In Southern Africa, Claire Janisch, one of biomimicry's pioneers in Africa, has been working with me and her colleagues to redesign housing slums in the Western Cape Province. She is transforming them into biologically inspired systems that manage water, deal with wastes, create soils, and modify local climates. I predict biomimicry will attract many others into the field of landscape restoration and carbon farming.

Another important person shaping global landscape transformation is filmmaker John Liu, a biologist specializing in large-scale land restoration. He has produced a film entitled *Green Gold* that illustrates large-scale restoration in China, which has been seen by more than a million people on YouTube. It is about re-greening a desert region known as the Loess Plateau. Twenty years ago the Chinese government set out to see if a desert could be turned into a biologically rich and diverse ecosystem. It was a vast barren region that stirred up dust that clouded the skies over large areas of China and caused Beijing's air quality to become almost unbreathable at certain times of year. The landscape became green through the terracing, planting, and caring efforts of large numbers of people. It now supports millions of people. The film has begun to influence European and Middle Eastern political leaders with its visible proof that deserts can become viable ecologies again.

John Liu is part of a group of Dutch and Belgian scientists and engineers who call themselves the Weather Makers. They have an interest in Northern Africa, the eastern Mediterranean, and the Middle East, specifically in the land bridge between Africa and Asia known as the Sinai Desert. Politically the Sinai is part of Egypt. The Weather Makers consider it to be a weather crucible that influences climate and weather far beyond its boundaries, impacting North Africa and the eastern Mediterranean with effects as far east as India and China.

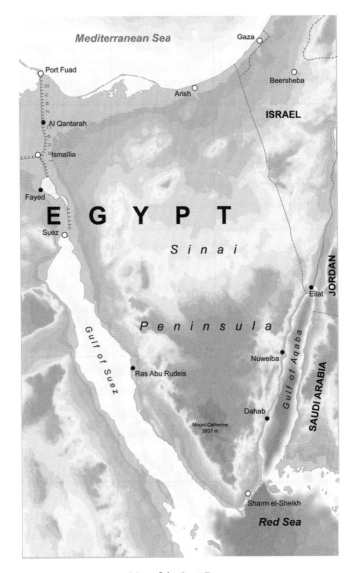

Map of the Sinai Desert

They argue that the Sinai was not always barren and devoid of vegetation; there was once a time when the Sinai's vegetation cooled the region, and helped bring in monsoon winds and with them the rains. The records of ancient seafarers seem to bear them out. Then over millennia the Sinai was deforested and overgrazed. This led to its soils being washed down from the mountains and hilltops towards the Mediterranean to the west. Most of these sediments ended up in Lake Bardawil, adjacent to the sea. Today its six hundred square kilometers are seriously silted up; the lake's depth is between three and six feet, whereas in Biblical times, it was between sixty and eighty feet deep. The Weather Makers see these accumulated ancient sediments as one source of organic materials for re-greening the Sinai.

The Weather Makers' theory, borne out by measurements, states that the Sinai is getting hotter and that the moisture in the atmosphere above it is lessening. The ability of the air to produce precipitation is declining and negatively influencing larger areas beyond the Sinai. To combat this, the Weather Makers intend to re-green the whole Sinai. With support from the government of the Netherlands, they have embarked upon an analysis of a variety of techniques for reforesting the region. They are emboldened in their task by the recent example of the re-greening of the large Loess Plateau in China, an example of a desert that has been botanically restored.

The Weather Makers invited me to assist them in devising technological and ecological solutions for beginning the restoration process. Their courage in the face of many obstacles and skeptics impressed me. Here was an opportunity to work at a country-sized scale with people deeply committed to the fate of the Earth.

I had in fact been thinking for many years about re-greening arid lands. I recall a conversation with the poet Gary Snyder in the mid 1970s while walking with him in our local woods near the sea on Cape Cod. We wondered if the Mediterranean could be restored to its condition in early classical times before the region was deforested and its biodiversity lost. Many species from that time are gone. Might enough species diversity be left to bring back the past? We realized that Earth has several regions with Mediterranean-like climates. Might it be possible for these regions to contribute species to fill the empty niches? Would an experiment, carried out on a biologically impoverished island in the Mediterranean, answer that question? I thought it an idea worth testing one day.

For the Weather Makers, I came up with an ecological engineering concept that uses seawater, an abundant resource along the coastal zones of the

Sinai Peninsula, as its central element. My proposal harkened back to my days at the New Alchemy Institute and my subsequent experience working with bio-shelters, solar structures that produced biological materials, cleaned wastes, and bountiful foods. Out of this experience I proposed a living technology I called Oasis eco-machines. Such structures could be used in large numbers to make freshwater and establish ecological and economic elements in the desert.

The main problem facing desert landscape restoration is lack of freshwater. I proposed to solve this by using geodesic greenhouse structures as climatic envelopes to regulate moisture in the air.

Within these geodesic structures, I suggested that we place tanks made of translucent materials to be filled with seawater. Such solar-aquatic systems are ideal for helping stabilize climate within the domes and for growing marine foods.

The next step is important. During hot sunny days, the dome vents are left wide open and air circulates freely, preventing overheating. At dusk the

Geodesic dome

vents are closed. During the night, as outside air cools, moisture accumulates on the interior ceiling and sides of the dome. After dawn, one of the staff comes along and drums with their hands on the sides of the dome. At this point the drumming vibrations cause it to rain inside the structure for several minutes. The dome, with its seawater tanks within, operates as a solar still, producing life-giving freshwater.

Solar aquatic tank

It now becomes possible to plant a wide variety of hardy perennial plants, shrubs, trees, and food crops inside the dome. The tanks serve double duty for the cultivation of seaweeds, clams, oysters, shrimp, and fish. The goal is to create an economic as well as ecological life support system in the Oasis eco-machines.

After several years, the systems would be ready to stand on their own. Deep roots will sustain the plants during the arid conditions to which they are increasingly exposed. At this juncture, the dome structure is moved to a new location and the cycle is repeated.

Dome with a fig tree and aquatic tanks within

I imagine a whole "fleet" of domes, hundreds or even thousands of them, slowly marching across the desert leaving bountiful living systems in their wake. Imagine the population of the area employed in tending them. It might be possible that in this way today's refugees from the region could find health and stability for their families. Through such steps, the planet could be re-greened, the climate stabilized, and all of humanity helped.

DRAWINGS OF AN OASIS ECO-MACHINE

This is a series of colored drawings based upon my designs illustrating the six stages of life emerging again in the Sinai Desert.

Climate envelope

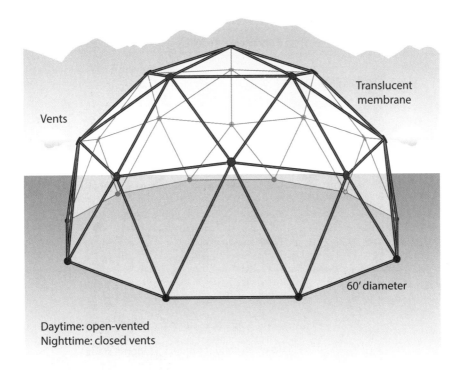

The structure

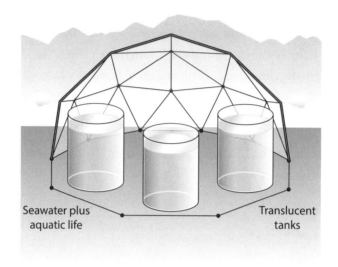

Seawater tanks within

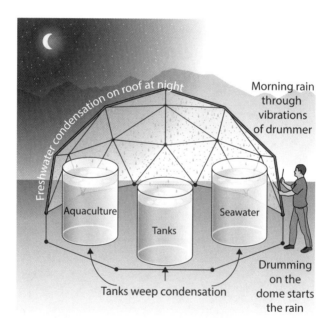

Freshwater rain at dawn

Re-Greening the Earth: The Challenge of the Sinai Desert 165

Cultivation of valuable seafoods

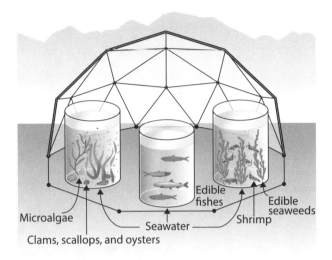

Saltwater farming

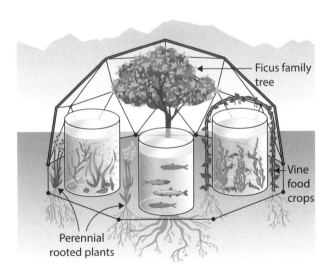

Establishing a terrestrial ecosystem within

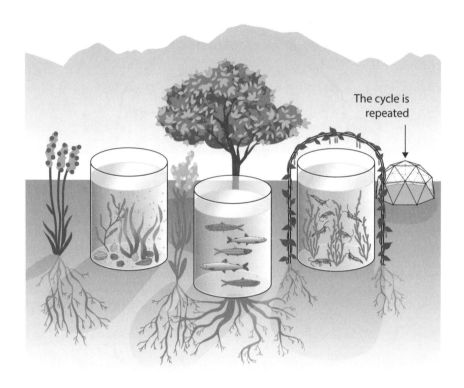

The dome is removed to a new location, leaving the ecosystem behind

I can foresee hundreds or even thousands of Oasis eco-machines producing food, freshwater, and ecological stability. Each proposed dome climate envelope is sixty feet in diameter. They can be built to be strong enough to withstand howling winds, sandstorms, and large fluctuations in temperature. Equally important, they can be built to be light; approximately twenty people can lift the structures and move them to new locations. The ecology that has grown up in them is left behind to continue to flourish.

The features of the Oasis eco-machines have already been tested elsewhere: saltwater farming in greenhouses, the ability of the structures to condense water out of the interior air at night, and their capacity to be botanically diverse and include trees such as figs.

Beyond the Oasis eco-machines, the Weather Makers have a whole host of ideas they want to explore on the Sinai. They want to dredge Lake Bardawil and connect it more actively to the Mediterranean. The lake, when flushed by Mediterranean waters, will be colonized by marine life including valuable sea foods. The dredged materials will be used throughout the Sinai as a soil base for upland restoration projects.

There is also an interest in growing halophytic or salt-tolerant plants; grains such as quinoa are included in a diversity of edible plants in saline areas. Also, there are specialized halophytes that can be used as ecological beachheads in the early stages of the larger restoration mission.

Another person whose pioneering ideas are influencing the Weather Makers as well as myself in the Sinai project is Dr. Carl Hodges, an atmospheric physicist, who founded the Environmental Research Laboratory at the University of Arizona in the late 1960s. A decade later he created the Sea Water Foundation to address the three issues of hunger and its relief, global warming, and rising sea levels. His primary research areas included halophyte plants, the cultivation of marine foods, and the uses for seawater in new and unusual ways. He focused on desert environments such as the Sonora region in Mexico and Egypt in the Middle East. I only knew him slightly from my New Alchemy Institute days, but was always impressed by his big picture holistic ideas and research.

At the Weather Makers' planning conference in the Netherlands, I came to realize that at least half a dozen of the people in the room worked for one of the world's largest dredging companies, Dredging International/Deme Group, headquartered in Belgium. I learned that these are the people who keep the Suez Canal dredged and open for the Egyptian government. They operate a fleet of more than ninety ships. One of the founders of the Weather Makers had been an engineer for the firm.

Afterwards, I began to make a connection between the dredging companies and several of Carl Hodges's ideas. He had come up with a totally unique approach to greening deserts that are adjacent to seacoasts. He proposed that long canals be dredged to form seawater rivers that would extend inland.

They would be at sea level and exchange ocean water during tides and storms. In the Netherlands, I learned that there were technologies and corporations that could build such canals and to do so over large areas. In this context, seawater could be another way of re-greening, adding a new and powerful dimension to Oasis eco-machines situated higher up on the slopes.

Carl Hodges proposed to plant very large numbers of mangroves, salt-tolerant trees normally found in estuaries and coastal lowlands in the tropics, in and adjacent to the canals—a plan that was equally relevant to our work in the Sinai coastline. Since some of the seawater will flow laterally from the canals into adjacent soils, the mangroves would spread with the saltwater. A forest would grow and provide useful biomass and timber on an ongoing basis. The trees themselves, through evapotranspiration, would reduce air temperatures and increase the moisture content in the atmosphere above.

Dr. Hodges also proposed irrigating with saltwater plants from the *Salicornia* family. This is a family of edible salt-tolerant succulents with a variety of names such as glasswort, pickle weed, or sea beans. They normally grow in mangroves, salt marshes, and coastal beaches. Many are nutritious and have a wide range of culinary uses. He recommended species of plants that cattle, sheep, and goats use as fodder. He wanted to create saltwater paddies along the canals for the mass propagation of *Salicornia*.

Carl Hodges was an exponent of the intensive culture of seafoods including shrimp and fish in the canals. The mangroves would host a diversity of marine life on their roots that are suitable as fish feed as well as protect young fish from predators. He suggested that such canals could help stabilize climate, produce energy, combat sea-level rise, and feed people, and I must say I agree with him that these measures are the way forward.

• • •

My hope for the future is predicated on very large numbers of people committing themselves to a life of Earth stewardship. Millions of such people will be needed to help stabilize climate and feed growing populations. These budding stewards will have to be trained to possess the skills needed for the ecological restoration of the planet. Today their numbers may be measured only in the thousands, or tens of thousands at best. The good news is that the knowledge and techniques are now widespread, and the even better news is that the large-scale training for such a task has begun.

As for the Sinai project, the global epicenter of training efforts in ecosystem restoration there is actually in the Netherlands, in an organization called Ecosystem Restoration Camps or ERC.[1] One of its leading lights is the filmmaker and soil scientist John Liu, mentioned earlier in relation to the Loess Plateau. The ERC mission is practical. They train people in restoration techniques and methods and restore degraded soil, and associated landscapes. The ERC organization is citizen-driven. It has built its first major camp or cooperative in a highly degraded part of Spain known as the Altiplano of Murcia. Although the region is depopulated, the Altiplano project works very closely with the remaining people to provide them with additional ecological support on their lands. Their long-term goal is to expand the camps and cooperatives worldwide. Already in place are associated projects in South Africa and Western Australia.[2]

It is the birth of projects like these that give me hope for the future. The possibility of a whole generation becoming environmentally literate and capable of engaging directly with the land is heartwarming. I have spent most of my life working with small, dedicated groups around issues of Earth stewardship. I feel very, very blessed to have been able to do so. But I look forward to the day when the world is truly green: when deserts are diminishing, forests growing, grasslands thriving, the inshore oceans are a cornucopia of healthy foods, and, most importantly, when human beings understand their role on Earth. There will be a time when carbon dioxide in the atmosphere is declining and climate is becoming more stable. The Earth is a bountiful place. This fact is my source of inspiration.

NOTES

CHAPTER 3: STEPS TO A THEORY OF ECOLOGICAL DESIGN

1. Howard T. Odum, *Environment, Power, and Society* (New York: Wiley Interscience, 1971).
2. John H. Todd and B. Josephson, "The Design of Living Technologies for Waste Treatment." *Ecological Engineering* 6, 1996, 109–136.
3. See http:// www.remineralize.org.
4. See Walter H. Adey and Karen Loveland, *Dynamic Aquaria: Building Living Ecosystems* (San Diego, CA: Academic Press, 1991); and John. H. Todd, E. J. G. Brown, and E. Wells, "Ecological Design Applied," in *Ecological Engineering* 20, 2003, 421–440.
5. Ocean Arks International, "Higher Plants at the South Burlington Advanced Ecologically Engineered System (AEES)," 2001. Unpublished document.
6. "A thirty year survey reveals the ecosystem function of fungi," *American Journal of Botany* March 98 (3), 2011, 426–438.
7. Paul Stamets, *Mycelium Running* (Berkeley, CA: Ten Speed Press, 2005).

CHAPTER 4: THE EDGE OF THE SEA

1. Walter Adey and Karen Loveland, *Dynamic Aquaria*, 2nd ed. (San Diego, CA: Academic Press, 1998).

CHAPTER 5: RESTORING POLLUTED WATERS ECOLOGICALLY

1. Walter Adey and Karen Loveland, *Dynamic Aquaria*, 2nd ed. (San Diego, CA: Academic Press, 1998).

CHAPTER 6: HEALING DEGRADED STREAMS AND RIVERS

1. J. A. Estes et al., "Trophic Downgrading of Planet Earth," *Science,* 333, 15 July 2011, 301–306.
2. B. J. Cardinale, "Biodiversity Improves Water Quality Through Niche Partitioning," *Nature* 472, 7 April 2011.

CHAPTER 15: APPALACHIAN SPRING

1. Acres USA (http://www.acresusa.com) is an important information source, and Urth Agriculture is a leading provider of microbial sprays (http://www.urthagriculture.com).
2. See http://www.rurdev.usda.gov/rbs/pub/cooprpts.htm.

CHAPTER 16: RE-GREENING THE EARTH: THE CHALLENGE OF THE SINAI DESERT

1. See http://www.ecosystemrestorationcamps.org.
2. See http://www.commonlandsfoundation.org.

BIBLIOGRAPHY

Bejan, Adrian, with J. Peder Zane, *Design in Nature.* New York: Anchor Books, 2013.

Benyus, Janine. *Biomimicry: Innovation Inspired by Nature.* New York: Harper Collins, 1997.

Falk, Ben. *The Resilient Farm and Homestead.* White River Junction, VT: Chelsea Green, 2013.

Leopold, Aldo. *A Sand County Almanac.* New York: Oxford University Press, 1949.

Mollison, Bill. *Permaculture: A Designer's Manual.* Tyalgum, Australia: Tagari Publications, 1988.

Mollison, Bill, and David Holmgren. *Permaculture One.* London: Corgi, 1978.

Savory, Alan. *Holistic Management*, 3rd ed. Washington, DC: Island Press, 2016.

Stamets, Paul. *Mycelium Running.* Berkeley, CA: Ten Speed Press, 2005.

Todd, Nancy Jack. *A Safe and Sustainable World: The Promise of Ecological Design.* Washington, DC: Island Press, 2005.

White, Courtney. *Grass, Soil, Hope.* White River Junction, VT: Chelsea Green, 2014.

ACKNOWLEDGMENTS

I want to express my heartfelt thanks to Richard Grossinger, the cofounder of North Atlantic Books. In 1993 he published my book *From Eco-Cities to Living Machines: Principles of Ecological Design,* written with my wife, Nancy Jack Todd. He was supportive and encouraging of this book, *Healing Earth.* Equally important, he conveyed his enthusiasm to his publishing colleagues at North Atlantic Books, most notably Pam Berkman and Tim McKee. I appreciate the careful work and help by the project editor, Louis Swaim, and the senior production artist, Emma Cofod. I would also like to thank Hisae Matsuda for her editorial assistance.

My life has been blessed with incredible coworkers, colleagues, professors, and students. There were numerous associates at the New Alchemy Institute in the 1970s and at Ocean Arks International from the early 1980s until the present. Both not-for-profit organizations hosted many of the innovations and experiments described in *Healing Earth.* Those people and institutions who supported the work have my deepest gratitude. The list includes, but is not limited to, the Rodale Organization, the Jessie Smith Noyes Foundation, the Rockefeller Brothers Fund, the National Science Foundation, the U.S. Environmental Protection Agency, the Massachusetts Center for Innovation in Marine and Polymer Sciences, the Mott Foundation, and the Lewis Foundation.

Along the way, companies were started to move the ideas into the marketplace. They included Four Elements Corporation, Ecological Engineering Associates, Living Technologies, and John Todd Ecological Design (JTED). Recently JTED has forged an alliance with Biohabitats, a national ecological design and environmental restoration firm.

Part of my life has been spent in academic institutions, including the Woods Hole Oceanographic Institution and the University of Vermont. My students expanded the scope and meaning of design and carried ecological concepts widely into their subsequent careers. The classrooms became a beacon of hope and innovation.

There are several individuals whom I must name. William O. McLarney, the cofounder with Nancy Jack Todd and me of New Alchemy. He is a brilliant ichthyologist and social ecologist. Margaret Mead, who urged Nancy and me to take the ideas of New Alchemy out into the world. We named our proposed ecological hope ship the *Margaret Mead* in her honor. The brilliant yacht designers Philip Bolger and Dick Newick both helped us envision working upon the ocean in different ways. Over the last few years, Laurie McGowan, a naval architect from Nova Scotia, has joined with us to develop the ocean restorer as a sailing ship and floating school. He is carrying on the legacy of assisting us with rethinking our relationships with the sea.

I wish to give special thanks to the illustrator of this book, Matt Beam, a former student, teaching assistant, and colleague. He combines an artistic sensibility with a profound knowledge of the systems he is illustrating.

Finally, I would like to acknowledge William Irwin Thompson, writer, cultural historian, and the founder and impresario of the Lindisfarne Association. Through Lindisfarne he has introduced us to some of the most seminal minds of our era.

INDEX

A

Adey, Walter, 36
agriculture, effects of, 44–45
algae
    blooms, 54, 80, 86
    blue-green, 22, 86
    brown, 23
    calcareous, 90
    green, 23
    red, 23
    turfs, 117–18
Altiplano project, 169
alum (aluminum potassium sulfate), 45–46
Andropogon gerardi, 144
animal kingdom, 16, 17, 26–28. *See also individual species*
Anodonta, 27
Appalachia, 137–54
Aqua-Forest concept, 57–58
Archaebacteria (kingdom), 16, 17, 22
Archer, Colin, 69
Atkin, William, 69
Atlantic coastal panic grass, 144–45
Australia, 156, 169
autotrophs, 22

B

bacteria
    important role of, 21–22
    kingdoms of, 22
    nitrification and, 67
Baima Canal, 107, 115
Bardach, John, 2
bass, large-mouth, 54
beach convolvulus, 90, 91, 94
beach erosion, 89–90
Beam, Matt, 83
Bejan, Adrian, 126–27
Benyus, Janine, 156
Berlin, Maryland, 81, 107, 114
Bernat, Eugene "Gino," 103–4, 107, 111
Bernat, Nick, 111
big bluestem, 144–45
biochar, 142, 146, 147, 148
biodiversity, 55, 156
Biology of Wastewater Treatment (Gray), 26
biomimicry, 156–57
Biomimicry: Innovation Inspired by Nature (Benyus), 156–57
Biomimicry of South Africa, 123
bio-shelters, 160
Blackstone River, 103
Bolger, Phil, 95–96
BP (British Petroleum) Deepwater Horizon disaster, 113
Brambilla, Roberto, 94
Branson, Sir Richard, 98
British Virgin Islands, 98
Bromfield, Louis, 52–53
browning, 43–44
Brown University, 110, 111
bryozoans, 31, 33, 65, 67
Buckminster Fuller Design Challenge, 141, 155
bulrush, 48
Bunker C oil, 26, 104–5
Burlington, Vermont, 39
Buzzards Bay, 31, 80

C

Canadian International Development Agency, 96
Canadian pondweed, 38–39
Cape Cod, 2, 3, 7, 31, 47, 61, 80, 82, 83, 91, 113, 159
Cape Town, South Africa, 121
Capper-Volstrand Act, 153
carbon
  dissolved organic, 43–44, 46
  farmers, 5
  sinks, 48
carcinogens, 7
Caribbean, 70, 87–94, 100
Carson, Rachel, 2
catfish, armored, 38
cellular design, 20–21
chemotrophs, 22
Chesapeake Bay, 79, 81
Chimaera, 99
China, 70, 81, 107, 115, 157, 159
Chlorophyta, 23
clams, 82
Clark University, 111
composting, 142
constructal law, 127
cooperatives, 152–53
coral reefs, 41, 54
Costa Rica, 19, 92, 97, 98, 141, 142
cyanobacteria, 22, 86

D

Dallas, Texas, 81–82
Daphnia, 27
DDT, 2, 104
deforestation, 44, 159
desert landscapes, re-greening, 155, 157–69
Design in Nature (Bejan and Zane), 126–27
Devil's Foot, 31
dinoflagellates, 23, 89
DOC (dissolved organic carbon), 43–44, 46
Doshi, Samir, 144
Dredging International/Deme Group, 167
Dynamic Aquaria (Adey and Loveland), 36

E

Ecological Aspects of Used Water Treatment (Curds and Hawkes), 26
ecological design
  diversity of life and, 35
  First Order of, 18, 29, 140
  guiding principles of, 19–29
  meta-intelligence and, 21
  nature as inspiration for, 9, 15, 126–27
  need for theory of, 16
  Second Order of, 18, 29, 140
  Third Order of, 18, 29, 140–41, 149
ecological fluidized beds, 47
ecological pipeline, 56, 57
eco-machines
  cascading (South African), 135
  definition of, 11
  designing, 18, 46–49, 70–71
  food-growing, 35–41, 76, 77
  micro-marine restorers, 83–86
  Oasis, 160–67, 168
  ocean restorers, 115–20
  oil decontamination, 104–5, 108–12
  pond restorers, 62–68, 81, 114, 115
  stream and river restorers, 56, 58–60
  waste treatment, 81, 114–15
eco-mimetics, 157
ecosystem assembly, 36
Ecosystem Restoration Camps (ERC), 169
ecosystems
  diversity and, 11
  nutrient reservoirs in, 19
  parent, 18, 21
  randomness and, 20
Edith Muma, 97
eelgrass
  description of, 71
  as ecosystem builder, 31–33, 72–74
  photographs of, 73, 74
  scientific classification of, 32, 71
  uses for, 71
eelgrass communities
  design inspired by, 34–41, 77, 117
  fish and, 31–32, 74–75
  vulnerability of, 33–34, 72, 77, 80
Egypt, 157, 167

Ehrlich, Karl, 63
elk, 54
Elodea canadensis, 38–39
Environmental Research Laboratory, 167
Environment, Power, and Society (Odum), 11
EPA. *See* U.S. Environmental Protection Agency
Eubacteria (kingdom), 16, 17, 22
Euglenophyta, 23
exchange rates, high, 20

F

Falk, Ben, 156
Falmouth, Massachusetts, 82
Farm Credit System, 153
feed conversion ratio (FCR), 40
fertilizers, 44, 142
First Order Ecological Design, 18, 29, 140
fish. *See also individual species*
    cultivating, 37–41, 149, 168
    in eelgrass communities, 31–32, 33, 34, 74–75
    as filter feeders, 27
Fisherville Mill, 103, 105
Flax Pond, 47, 61–68, 113–14
floating parks, restorers as, 48, 49, 82
fungi
    kingdom of, 16, 17, 25–26
    mycorrhizal, 16, 17
Fuzhou, China, 107, 115
fynbos, 122, 123

G

Gaia (Lovelock), 4
Gaia theory, 4
geodesic domes, 160
glomalin, 45, 46–47
Golding, William, 89
gradients, steep, 20
Grafton, Massachusetts, 103, 105, 107–8, 110
Grass, Soil, Hope (White), 44
Gray, N. F., 26
Green Gold (film), 157
Gulf of Mexico, 79, 81, 113, 114, 115, 120
Guyana, 92, 96, 97

H

halophytes, 167
Hamilton Bay, 51
Harwich, Massachusetts, 7, 61, 113
Healing Gaia (Lovelock), 4
herbicides, 44
heterotrophs, 22
Heyerdahl, Thor, 120
Hodges, Carl, 167–68
Holistic Management (Savory), 155
Holmgren, David, 156
humus, creation of, 45

I

In-formal South, 123
Intervale, 39
Irens, Nigel, 93

J

Janisch, Claire, 157
JTED design, 70

K

kelp forests
    artificial, 56, 57–58, 117
    healthy, 53–54
Koi, 28
Kona, Hawaii, 81, 108, 115

L

Labrynthula zosteracea, 34
Laird, Marshal, 1
Lake Bardawil, 167
Lake Ontario, 51, 69
landscape management, effects of, 44–45
land trusts, 151, 152
Langrug, South Africa, 123–25, 128
Lebdig's law, 46
Leopold, Aldo, 155
Lewis Foundation, 137, 141
life, kingdoms of, 16, 17
Liu, John, 157, 169
Living Systems Laboratory, 111
Loess Plateau, 4, 157, 159
Lord of the Flies (film), 89

Loveland, Karen, 36
Lovelock, James, 4
luciferin, 89
Lymnaeidae, 27

M

macrocosms, 28–29
Malabar Farm (Bromfield), 52–53
malaria, 15, 23
Maluti Waters, 123
Mandela, Nelson, 121
mangroves, 41, 89, 168
Margaret Mead, 95
Margulis, Lynn, 3–4, 11, 21
Marine Biological Laboratory, 11
Martha's Vineyard, 69, 91, 104
Massachusetts Department of Environmental Protection, 12, 110
Massachusetts Foundation for Excellence in Marine and Polymer Sciences, 12
Mbekweni, South Africa, 123
McGill University, 1
McGowan, Laurie, 99, 100, 118
McLarney, Bill, 3
mesocosms, 28
meta-intelligence, 21
methanogens, 22
Mexico, 167
microcosms, 28
micro-marine restorers, 83–86
mineral diversity, 19
mining
    environmental impact of, 137–38
    restoration after, 141–54
Mississippi River, 113, 120
Mollison, Bill, 156
Mthambo Development Services, 123
mushrooms, 25, 26, 104
mussels, 82
Mycelium Running (Stamets), 25
mycorrhiza, 45
Myxomycetes, 23

N

Nancy Jack, 96

Nantucket Sound, 31, 69, 104
nature
    ecological information within, 15
    law of design in, 126–27
    self-repair ability of, 17
    as source of inspiration, 9, 15, 71, 126–27
Netherlands, 159, 167, 168, 169
New Alchemy Institute, 3, 62, 91, 160, 167
Newick, Dick, 91, 96–97, 99
niche diversity, 55
nitrification, 67
nutrient reservoirs, 19

O

Oasis eco-machines, 160–67, 168
ocean
    pollution, 80
    restorers, 115–20
Ocean Arks International, 70, 93, 95–96, 97, 99
Ocean Pickup, 91–92, 96, 97, 99
Odum, Howard T., 11, 16
oil
    Bunker C, 26, 104–5
    spills, 79, 113, 115
Oomycetes, 23
otters, 53
oysters, 82

P

Panicum amarum, 144
P. virgatum, 144
parent ecosystems, 18, 21
Pelicano, 99–102
permaculture, 156
Permaculture: A Designer's Manual (Mollison), 156
Permaculture One (Mollison and Holmgren), 156
pesticides, 2, 44, 104
Phaeophyta, 23
Phosphorescent Bay, 89
photosynthesis, 22, 23, 24
phototrophic soil makers, 131–33, 135

Physa gyrina, 105
Physidae, 27
pipefishes, 75
Plankenburg River, 55, 123
plant kingdom, 16, 17, 24–25. *See also individual species*
Plasmodium vivax, 23
polluted waters
    causes of, 43–44, 80
    contemporary methods of remediating, 45–46
    eelgrass communities and, 34
    effects of, 43, 61, 79–80
    responsibility and, 80
    restoring ecologically, 46–49, 55–68, 114–20
Powell River Project, 143, 144
proas, 97
Protista (kingdom), 16, 17, 22–24
Protozoa, 23
Providence, Rhode Island, 80, 103, 110
Puerto Rico, 87, 88

R

randomness, benefits of, 20
Red Brook Harbor, 83, 85
reforestation, 143–44
Reilly, William, 12
The Resilient Farm and Homestead (Falk), 156
Rhodophyta, 23
rivers
    decline in, 54–55
    restoring, 55–60
rock powders, 19, 141–42
Rome, Max, 133

S

A Safe and Sustainable World (Todd), 3
Salicornia, 168
A Sand County Almanac (Leopold), 155
San Diego State University, 2
Savory, Alan, 155–56
scallops, bay, 33, 34, 73, 74, 77, 80
Schacht, Michael, 99

Scirpus spp., 48
seahorses, 75
Sea Water Foundation, 167
Second Order Ecological Design, 18, 29, 140
septic tank waste, 7–8
Seychelles, 70
sharks, 54
shellfish culture, 82–83
shrimp aquaculture eco-machine, 76, 77
Silent Spring (Carson), 2
Sinai Desert, 4, 155, 157–67, 169
slime molds, 23, 34
snails, 27, 105
Snyder, Gary, 159
soil
    healthy, 45
    loss of, 44, 141
    makers, phototrophic, 131–33, 135
    rebuilding, 44–45, 53, 141–46
solar power, 9–12, 56, 58–60, 99, 119
South Africa, 55, 59, 60, 121–36, 157, 169
South Burlington, Vermont, 24
Spain, 169
Stamets, Paul, 25
Stellenbosch, South Africa, 121, 125
sticklebacks, 75
St. Lawrence River, 69
streams
    decline in, 54–55
    differences in, 51–52
    restoring, 55–60
succession
    cultural, 149–53
    ecological, 29, 140–41, 143–44
Suez Canal, 167
switch grass, 144–45
Symbiosis in Cell Evolution (Margulis), 4

T

Table Mountain, 121
Tahiti, 70
thermodynamics, 126, 127
Third Order Ecological Design, 18, 29, 140–41, 149

tidal upwellers, 117
tilapia, 37, 41
TOC (Total Organic Carbon), 105
Todd, Nancy Jack, 2, 3, 121
TPH (Total Petroleum Hydrocarbon), 105
tree wells, 131–33
Trinity River, 82
trophic downgrading, 53–54, 56
Tufts University, 111

U

Unio, 27
University of Arizona, 167
University of Costa Rica, 97
University of Michigan, 2, 3
University of Vermont, 144
Urth Agriculture, 142
U.S. Environmental Protection Agency, 7, 12–13, 105, 110, 111

V

Vieques, 87–94
Virginia Tech, 143
VOCs (volatile organic compounds), 61

W

waste treatment, 7–13, 26–29, 81, 125–36
water. *See also* polluted waters
  importance of, 70
  quality, 70
water molds, 23
Weather Makers, 155, 157–59, 167
White, Courtney, 44
wind power, 62–68, 96, 99, 119
wolves, 54
Woods Hole, Massachusetts, 11, 31
Woods Hole Oceanographic Institution, 2, 3, 35, 104
Worcester, Massachusetts, 103
Worcester Polytech, 111

Y

yeasts, 26

Z

Zane, J. Peder, 126
Zosteraceae, 32
Zostera marina. See eelgrass

ABOUT THE AUTHOR

One of the pioneers in ecological design, John Todd has been active in shaping the discipline for over forty years. He has befriended and inspired such well-known names in the field as Janine Benyus, Bill McKibben, Paul Hawken, David Orr, and Wes Jackson.

Together with his wife, Nancy Jack Todd, John Todd was named as a visionary of the twentieth century by the British organization Resurgence, and he has been featured in "The Genius Issue" of *Esquire* magazine. Dr. Todd was also named one of the twentieth century's top thirty-five inventors by the Lemelson-MIT Program for Invention and Innovation.

In 2010 the Smithsonian Institution recognized his eco-machine at the Center for Sustainable Living, part of the Omega Institute in Rhinebeck, New York, as a significant design breakthrough. The drawings and film of the project were displayed in the Smithsonian's Cooper-Hewitt Museum of Design in New York City.

About North Atlantic Books

North Atlantic Books (NAB) is an independent, nonprofit publisher committed to a bold exploration of the relationships between mind, body, spirit, and nature. Founded in 1974, NAB aims to nurture a holistic view of the arts, sciences, humanities, and healing. To make a donation or to learn more about our books, authors, events, and newsletter, please visit www.northatlanticbooks.com.

North Atlantic Books is the publishing arm of the Society for the Study of Native Arts and Sciences, a 501(c)(3) nonprofit educational organization that promotes cross-cultural perspectives linking scientific, social, and artistic fields. To learn how you can support us, please visit our website.

21982318962887